Genome, Transcriptome and Proteome Analysis

Genome, Transcriptome and Proteome Analysis

Alain Bernot

Généthon III, France

Translated by

James McClellan
and **Susan Cure**

John Wiley & Sons, Ltd

First published in French as *Analyse de Génomes, Transcriptome et Protéoms* © 2001 Dunod, Paris

Translated into English by James McClellan with excerpts translated by Susan Cure

This work has been published with the help of the French Ministère de la Culture-Centre national du livre

English language translation copyright © 2004 by John Wiley & Sons Ltd,
The Atrium, Southern Gate,
Chichester, West Sussex PO19 8SQ,
England

Telephone (+44) 1243 779777

Email (for orders and customer service enquiries): cs-books@wiley.co.uk
Visit our Home Page on www.wileyeurope.com or www.wiley.com

Reprinted September 2005

Other Wiley Editorial Offices

John Wiley & Sons Inc., 111 River Street, Hoboken, NJ 07030, USA

Jossey-Bass, 989 Market Street, San Francisco, CA 94103-1741, USA

Wiley-VCH Verlag GmbH, Boschstr. 12, D-69469 Weinheim, Germany

John Wiley & Sons Australia Ltd, 33 Park Road, Milton, Queensland 4064, Australia

John Wiley & Sons (Asia) Pte Ltd, 2 Clementi Loop # 02–01, Jin Xing Distripark, Singapore 129809

John Wiley & Sons Canada Ltd, 22 Worcester Road, Etobicoke, Ontario, Canada M9W 1L1

British Library Cataloguing in Publication Data

A catalogue record for this book is available from the British Library

ISBN 10: 0 470 84954 1 (HB) ISBN 13: 978 0 470 84954 5 (HB)
ISBN 10: 0 470 84955 X (PB) ISBN 13: 978 0 470 84955 2 (PB)

Typeset in 10.5 on 13 pt by Kolam Informations Services Pvt. Ltd, Pondicherry, India

Contents

Preface

The Genome Project began in 1990. This project's aim is the analysis of the human genome, along with those of model organisms, to determine the location of all the genes in those genomes, and finally to establish their complete sequence. This objective is at the same time very simple, because a genome can be entirely described by the order of four letters A, T, G and C, and very complex, because the human genome contains more than three thousand million such letters.

This project is the first enterprise of truly international stature in the biomedical field. Because of its exceptional ambition it has been compared to the Apollo programme in space travel, to the quest for the Holy Grail, or to the establishment of the Periodic Table for biology. The completion of this work will greatly advance our knowledge in the areas of both biology and health.

The Genome Project should put at our disposal the necessary tools to understand and to treat numerous diseases having a degree of genetic predisposition. The data produced will also be of considerable interest for fundamental biology. The first results of the sequencing of model organisms already give a foretaste of this. As always in research, the results lead to at least as many questions as answers.

Since 1990, the development of the genome programmes has led to the establishment of genetic and physical maps, and the sequencing of the genomes of model organisms (or pathogens) and man. Complete sequencing of the genomes of representatives of the major groups of life has been achieved, and that of man, whilst currently partial, will probably be finished in 2003. In parallel, programmes for the analysis of transcription (transcriptome) and of translation (proteome) have been developed both in model organisms and man.

This book summarizes the work already achieved, and anticipates the likely scale of the discoveries to come, with their implications for fundamental biology and medicine. It gives the most up-to-date synthesis of these scientific enterprises.

Alain Bernot

About the Author

Alain Bernot is a graduate of the *École normale supérieure*, PhD, and Professor at the Université d'Evry. He is currently working in a genetic therapy programme at Genethon. He previously directed the sequencing and analysis of a vertebrate genome (*Tetraodon nigroviridis*), and contributed to the identification of a gene responsible for a human genetic disease (Mediterranean familial fever).

Acknowledgements

Figure 2.4 – from A genetic linkage map of the human genome, Donis-Keller H. *et al*, ©*Cell*. 1987; A second-generation linkage map of the human genome, Weissenbach J. *et al*, ©*Nature*. 1992; The 1993–94 Genethon human genetic linkage map, Gyapay G. *et al*, ©*Nat Genet*. 1994; A comprehensive genetic map of the human genome based on 5,264 microsatellites, Dib C. *et al*, ©*Nature*. 1996.

Figure 2.5 – from Single-Nucleotide Polymorphism Identification Assays Using a Thermostable DNA Polymerase and Delayed Extraction MALDI-TOF Mass Spectrometry, Haff LA & Smirnov IP, ©*Genome Res*. 1997.

Figure 3.8 – from Continuum of overlapping clones spanning the entire human chromosome 21q, Chumakov I. *et al*, ©*Nature*. 1992.

Figure 4.3 – from Whole-genome random sequencing and assembly of *Haemophilus influenzae* Rd, Fleischmann RD. *et al*, ©*Science*. 1995.

Figure 4.4 – from Analysis of the *Escherichia coli* genome VI: DNA sequence of the region from 92.8 through 100 minutes, Burland V. *et al*, ©*Nucleic Acids Res*. 1995.

Figure 4.5 – from Genomic sequence of a Lyme disease spirochaete, *Borrelia burgdorferi*, Fraser CM *et al*, ©*Nature*. 1997.

Figure 4.6 – from The complete genome sequence of the gram-positive bacterium *Bacillus Subtilis*, Kunst F *et al*, ©*Nature*. 1997.

Figure 4.8 – from Genomic-sequence comparison of two unrelated isolates of the human gastric pathogen *Helicobacter pylori*, Alm, RA. *et al*, ©*Nature* 1999.

Figure 4.9 – from The *C. elegans* Sequencing Consortium. Genome sequence of the nematode C. elegans: a platform for investigating biology, ©*Science* 1998.

Figure 4.10 – from The DNA sequence of human chromosome 22, Dunham I. *et al*, ©*Nature*. 1999.

Figure 4.12 – from The genome sequence of the malaria mosquito *Anopheles gambiae*, Holt RA *et al*, ©*Science*. 2002.

Figure 4.15 – from Recent developments from the *Leishmania* genome project, Myler PJ & Stuart KD, ©*Curr Opin Microbiol*. 2000.

Figure 5.6 – from A DNA microarray system for analyzing complex DNA samples using two-color fluorescent probe hybridization, ©*Genome Research* 1996.

Figure 5.7 – from Molecular portraits of human breast tumours, Perou CM *et al*, ©*Nature*. 2000.

Figure 5.9 – from Accessing Genetic Information with High-density DNA arrays, Chee M. *et al*, ©*Science*. 1996.

Figure 6.12 – from A two-dimensional protein map of Chinese hamster ovary cells, Champion KM *et al*, ©*Electrophoresis*. 1999

Figure 6.17 – from Encyclopedia of Molecular Biology and Molecular Medicie, Robert A. Raymon E. Kaiser Jr, VCH.

Figure 6.19 – from Novel interactions of *Saccharomyces cerevisiae* type 1 protein phosphatase identified by single-step affinity purification and mass spectrometry, Walsh EP *et al*, ©*Biochemistry*. 2002.

Figure 6.25 – from The foreign antigen binding site and T cell recognition regions of class I histocompatability antigens, Bjorkman PJ. *et al*, ©*Nature*. 1987.

1 General Introduction

The genome of a living thing is made from DNA, a molecule which contains within its linear sequence of nucleotides all the information comprising the biochemistry of life. The functional unit of the genome is the gene. Each cellular protein is the product of the expression of a gene, and the assembly of these proteins is the basis for the construction and maintenance of the cells of the organism.

For any given species, the genome is globally similar in all individuals. Nevertheless there are variations between the genomes of different individuals; this is known as polymorphism. Polymorphism can be silent, or can manifest itself by modifying an observable characteristic. Sometimes it can induce a serious dysfunction of the organism (a genetic disease).

The analysis of whole genomes makes use of the tools of molecular biology. The change of scale in comparison to the studies tackled by traditional laboratories has nevertheless led to the development of more specialized aspects, both at the technical and organizational levels. Before examining these, we present a brief review of molecular genetics.

1.1 Review of Molecular Genetics

1.1.1 Nucleic acids

DNA (desoxyribonucleic acid) is a molecule made by the linear assembly of four nucleotides: deoxyadenosine-, deoxycytidine-, deoxyguanosine- or deoxythymidine-phosphate. Each is composed of a nitrogenous base (respectively adenine, cytosine, guanine or thymidine), a sugar (deoxyribose) and a phosphate group. It is customary to use the simple letters A, C, G and T to describe the sequence of DNA. RNA

Genome, Transcriptome and Proteome Analysis by Alain Bernot
© 2004 John Wiley & Sons, Ltd   ISBN 0 470 84954 1 (HB)   ISBN 0 470 84955 X (pbk)

(ribonucleic acid) is also a linear molecule comprising four types of nucleotides, but here the sugar is ribose, and uracil (U) replaces thymidine. Because of the asymmetry of the sugar, these molecules are oriented, and the connection of the nucleotides is described in terms of the numbering of the carbon atoms in the sugar (orientation $5'-3'$, or $3'-5'$).

In vivo, DNA is practically never found except in the double-strand form, where the orientation of each strand is opposite. The strands form a double helix at the heart of which the nitrogenous bases of one strand are held face-to-face with the bases of the other strand by hydrogen bonds. The two strands of a molecule of DNA are complementary (Figure 1.1).

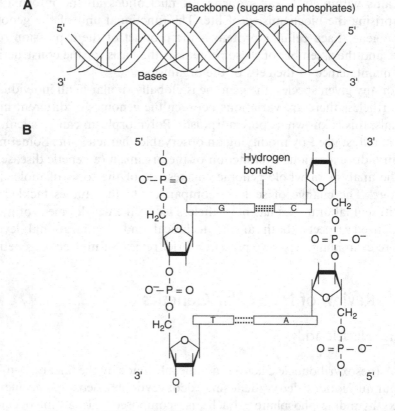

Figure 1.1 Structure of DNA. (A) Three-dimensional structure of the DNA helix. A $5'-3'$-oriented strand is always paired to a $3'-5'$-oriented strand. (B) Arrangement of the phosphate groups and desoxyriboses which constitute the two helical strands between which the nitrogenous bases interact. A cytosine (C) is always paired to a guanine (G), and an adenine (A) to a thymine (T). The two chains are oriented according to the $5'-3'$ direction of the sugars

Experimentally, two complementary strands can be separated – a process termed denaturation – by the action of heat or chemical agents for example. Under certain favourable conditions, the complementary strands may re-anneal: this is termed hybridization. The possibility of reversible interconversion between the denatured and renatured states is very important for several of the techniques summarized in this chapter.

1.1.2 Constituents of the genome

DNA is the store of genetic information in practically all living things (only a few viruses have RNA genomes). This information is represented on the chromosome by a linear suite of genes, separated by intergenic regions. For each gene, one of the two strands of DNA is transcribed into a complementary molecule of RNA. The messenger RNAs are translated into proteins. Transfer RNAs (tRNA) and ribosomal RNAs (rRNA) are used during the stage of the translation of messenger RNA into protein.

In prokaryotes (bacteria), the genome generally comprises a single circular chromosome. Coexisting with this may be extrachromosomal structures, in the form of generally circular molecules of DNA, which replicate in an independent manner: these are known as plasmids. Through the science of genetics bacterial cloning vectors have been derived from certain of these structures.

Eukaryotes are characterized by the existence of a nucleus, within which is the genome. This comprises chromosomes, the number of which is constant for a given species. Each chromosome is a long un-interrupted molecule of DNA, complexed with various proteins. Chromosomes generally have two arms separated by a primary constriction, which contains the centromere. Through their centromeres chromosomes are connected with the spindle fibres during cellular division. If the centromere is at the middle of a chromosome, the chromosome is said to be metacentric. It is sub-metacentric if the two arms are of unequal length, and acrocentric if the centromere is situated at the end of the chromosome. The terminal regions of chromosomes are called telomeres.

Most of the eukaryotic organisms discussed in this work are diploid, which means that their chromosomes are present in pairs: several pairs of homologous chromosomes (autosomes), and two sex chromosomes (but there is no sex chromosome in *Arabidopsis thaliana*, and yeast can replicate in the haploid as well as the diploid state).

In diploid organisms, each locus (a locus can be a gene, or any other position chosen on a chromosome) is present in two copies, called alleles. At a given locus, the two alleles may be identical, or may exhibit subtle differences: the locus is described as homozygous in the first case, heterozygous in the second.

During meiosis, the two chromosomes of each pair are separated. Each gamete formed possesses only one copy of each chromosome, which corresponds to the haploid state. Fertilization re-establishes the original number of chromosomes.

The human genome comprises 22 pairs of autosomes and one pair of sex chromosomes: XX in females and XY in males. The physical length of the total haploid genome is about 3000 million base pairs (a million base pairs is a megabase, abbreviated Mb).

1.1.3   Organelle genomes

As well as the nuclear genome, the organelles of eukaryotic cells have their own separate genome. These organelles are the mitochondria (organelles which mediate cellular respiration), and the chloroplasts of plant cells (organelles which mediate photosynthesis). Generally their genomes comprise a circular molecule of DNA.

The mitochondrial genome has been completely sequenced in several species. In humans, it comprises 37 genes spread over 166 000 base pairs (a thousand base pairs is a kilobase, abbreviated kb). Mutations in this genome are responsible for several genetic diseases (mitochondrial disorders). They are identifiable because they are entirely maternally inherited, owing to the fact that mitochondria are only inherited from the mother (in mammals).

Chloroplasts constitute a fundamental characteristic which distinguishes plant cells, autotrophs, from other eukaryotic cells, heterotrophs. The chloroplast genome of angiosperms frequently comprises between 120 and 160 kb. That of *Arabidopsis thaliana* covers 154 kb and contains 87 genes.

The size of the organelle genome is very modest by comparison with that of the nuclear genome. From the informational point of view, the preponderance of the latter over the former is overwhelming.

1.1.4 The structure of genes

In prokaryotes, the sequence of each gene is continuous, but in eukaryotes, a gene is most often composed of exons separated by introns. The exons are the parts of the gene which are preserved in the mature RNA, while the introns are eliminated (Figure 1.2). Exons contain the sequence which is translated into protein: they comprise the coding sequence. Generally, only the beginning of the first exon and the end of the last exon are not translated.

The regulation of gene expression is mediated by special sequences: the promoters (universal among living organisms) situated immediately upstream of the site of transcriptional initiation, and enhancers and silencers (elements that modulate gene expression, specific to eukaryotes), which can occupy very diverse locations with respect to the gene they regulate.

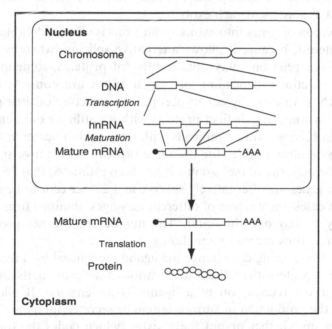

Figure 1.2 Transcription and maturation of messenger RNAs in eukaryotes. The whole gene (comprising three exons in this example) is transcribed in the form of a messenger RNA precursor (hn RNA). Transcription and maturation take place in the nucleus. In the cytoplasm, the mRNA is translated into protein

In humans, the average size of genes is of the order of 30 kb, but there exist genes covering only 400 bp, and very large genes covering more than 1 Mb (the gene coding dystrophin extends over 2.4 Mb).

1.1.5 Transcription and translation

Transcription is the synthesis of a molecule of RNA complementary to a sequence of DNA. Messenger RNAs are destined to be translated into proteins in the cytoplasm; tRNAs and rRNAs are implicated in the phenomenon of translation.

Messenger RNAs are translated without modification in prokaryotes. In eukaryotes, however, pretranslational modification of messenger RNA is extensive. In the first place, the ends are modified: the 5′ end by a cap, the 3′ end by the addition of a poly-adenylated tail. Additionally, the intron sequences are eliminated and the exon sequences joined up. This process is called splicing. Splicing requires specific sites around the 5′ and 3′ borders of each exon.

The division of genes into exons and introns is of considerable physiological interest, because it allows alternative splicing, whereby starting from a given gene one may obtain different proteins, sometimes with different functions, depending on which exons are conserved in the mature RNA. In this way eukaryotes put the structure of their genes to good use, synthesizing distinct proteins without utilizing different genes. These possibilities contrast radically with the world of bacteria, in which a given gene always codes for the same protein. Alternative splicing is used by the majority of metazoans. It has been estimated that 35 per cent of human genes are alternatively spliced, and 22 per cent in nematodes. In certain cases, the number of different messages obtained from a single gene may exceed one thousand. The number of distinct proteins in humans may thus exceed one million.

Alternative splicing can change the ligand recognized by a receptor or an adhesion protein, the cellular localization of the protein, its phosphorylation, or the recognition of a ligand by an enzyme. It allows, for example, the production of surface-bound or secreted immunoglobulins, depending on whether or not the 3′ exon (which codes the transmembrane region) is preserved in the mature RNA. In the same way acetylcholine-esterase can be obtained in a cytoplasmic form, joined to the membrane by a glycophospholipid association, or it may be secreted. In some cases, the functions of two proteins produced by alternative

A

```
.......GCTGAGCCGGCTCCTGAGAGAAGCGCTTTCTGAGTCGTTTCGAG
GACAGCCCTGGCCGGTCTTTCCAGGCTGTGAGGGGCTCCTGGGACTGCTGT
CTCCTCTTATCCTGTACCTCTGCCATGTGACACACACACACACACACACAC
ACACACACACACACACACACACACACACATAAATTATCCTGGAGGAAAGGT
TAAGGTGACACATGGAGACTGAGTGTCACCGTTATTTCCGCAGGTCCTCTC
TGATGACATGAAGAAGCTGAAGGCCCGAATGGTAATGCTCCTCCCTACTTC
TGCTCAGGGGTTGGGGGCCTGGGTCTCAGCGTGTGACACTGAGGACACTGT
GGGACACCTGGGACCCTGGAGGGACAAGGATCCGGCCCTT.....
```

B

```
.......TCAGGGTGAGAAGGATGAAAACGGGACCCACAGGCTCCCTCACCC
CTTACCGTGGGCAAATGCTTGCACCTGGGTGGCAGTGAGTGGGCGGGTAAT
CGGGCAGGAGGGGGAGGCGGGCAGGAGGGGGGAGGCGGGCACCAGGGGCGA
GGTGAGCAGGAGGGGGAGGCGGGCACGAGGAGGAGGCGGGCAGGAGGGGGA
GACGGGCAGGAGAGGGAGGCAGGCAGGAGAGGGAGGTGGGCAGGAGGGGGG
GGCGGGCAGGAGGAGGAGGTGGGCAGGAGGGGGAGGCGGGCAGGAGGGGGA
GGCGGGCAGGAGGGTGAGGGGGGATCTGGACGCCCGGGGAGACTGAGGGAG
GCATCCAAGCCCCAGGGCTCCTTGAGGAAACAACAGGGGTGCCAGACGTGG
CCCGGGCCCCTGGCTGGGCCCAGTTCGGGGTGTGTGGGAGCTGAGGACTCA
CTGGGCTTGAGGACTGACTGATGTGGA.....
```

Figure 1.3   Examples of satellite DNA sequences (underlined). (A) Microsatellite: the size of the repeated motif can vary between one and about ten of nucleotides. The microsatellite represented here is a repeat of 25 CA dinucleotides. (B) Minisatellite: the size of the repeated motif can range from tens to hundreds of nucleotides. Here the motif is the sequence CGGGCAGGAGGGGGAGG (whose repetition is not strictly identical from one motif to another)

splicing are completely distinct: the same gene can, for example, code calcitonin in the thyroid, or calcitonin gene-related peptide in the brain (CGRP, a neuromediator).

The most impressive example is found in the gene for DSCAM of *Drosophila*. In the case of this gene, regions may be represented by a large number of different exons: for example, there are 12 different possible fourth exons, 48 sixth exons, 33 ninth exons and two seventeenth exons. Consequently, 38 016 different mature RNAs may be derived from the primary transcript of this gene.

The translation of RNA messengers into protein is mediated by a complex machinery comprising numerous proteins, rRNA and tRNA. Translation is effected through the genetic code, whereby each triplet of DNA bases (codon) is translated into an unique amino acid. Messenger RNA is translated between an initiation codon and a termination codon, this interval defining the open reading frame (ORF). The eukaryote initiation codon AUG is also predominant in prokaryotes, but other initiation codons are used (for example in *E. coli* 14 per cent of initiation codons are GTG, and 3 per cent are TTG); the termination codon is UAG, UAA or UGA. The finally synthesized protein is strictly colinear with the mature messenger RNA, its amino terminal corresponding to the 5′ end of the messenger RNA and its carboxy-terminal to the 3′ end of the mRNA.

Proteins frequently undergo post-translational modification, and there are many different possible modifications: these may include glycosylation, phosphorylation, prenylation/isoprenylation, acylation, deamination or ubiquitination. The same gene can thus give rise to a large number of distinct proteins.

Finally, besides rRNA and tRNA, other genes encoding non-translated RNAs have been identified. For several of these a role has been identified: some are implicated in the regulation of transcription or translation, some in the modification of tRNA, rRNA, mRNA, some in the maintenance of telomeres, etc. The gene XIST, for example, is found on the X chromosome. Its transcript of 16.5 kb contains no ORF and thus is not translated, but rather is implicated in the transcriptional inhibition of practically all the genes on one of the two X chromosomes in all the cells of female mammals (whose karyotype contains two X chromosomes).

1.1.6   Genes, multigene families and conserved sequences

Many genes are represented only once in the haploid genome. There are also cases of duplication, where the same gene is present in several copies. This is the case, for example, in the genes for globins (several copies per genome) or the genes for rRNA (several hundred copies).

Some genes belong to multigene families: the sequences of these genes are very similar to one another, and they encode similar proteins. Such cases are frequent: they include the genes encoding histocompatibility molecules, odour receptors and actins.

Finally, it is possible to identify sequences conserved between many genes, which encode putative functional domains. Even though the conservation of these sequences may be weaker than in the case of multigene families, informational tools for sequence comparison allow the identification of a large number of domains, which are found in numerous genes, e.g. immunoglobulin, fibronectin, lectin and homeo-box domains. A group of genes containing similar domains constitutes a superfamily. The same gene may also contain several different domains.

1.1.7 Coding and non-coding fractions of the genome

In prokaryotes, virtually all of the genome is coding: more than 85% of the bacterial chromosome is transcribed into RNA. In contrast, eukaryote genomes often contain only small amounts of coding DNA. In mammals it is estimated that this fraction represents less than 5 per cent of the genome. The non-coding fraction of DNA contains gene-regulatory regions, but also a large number of sequences having no known function.

The non-coding fraction includes unique sequences, tandemly repeated sequences, and dispersed repeated sequences. Tandemly repeated sequences include microsatellites and minisatellites, which are classified according to the size of the repeat (Figure 1.3). Dispersed repeated sequences are conserved sequences spread throughout the genome. In humans, they include SINEs (short interspersed elements) and LINEs (long interspersed elements), respectively represented by about 1.5 million and 850 000 examples per haploid genome (cf. section 4.3).

1.2 The Tools of Molecular Biology

1.2.1 Restriction enzymes and electrophoresis

Restriction enzymes are capable of specifically recognizing a short sequence, from 4 to 10 bp, and cleaving the DNA at the recognized site. They allow the fractionation of DNA into segments of reduced size, or to cut it at one or another desired site. Several hundreds of these enzymes have been characterized (e.g. *Eco*RI, *Bam*HI, *Not*I), and they recognize a wide variety of cut sites.

Restriction enzymes may be used to establish restriction maps of any molecule of DNA which one wishes to characterize. This consists of determining the order of restriction sites along the molecule, using electrophoresis to determine the sizes of the fragments liberated (Figure 1.4). Although this method of characterization can be frustrating, it is really

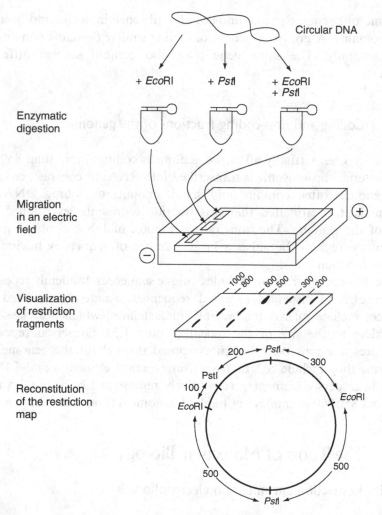

Figure 1.4 Restriction map of a DNA fragment. Digestion by a restriction enzyme produces a collection of fragments, whose size depends partly on the DNA sequence, and partly on the site recognized by the enzyme. The sizes of the DNA fragments are determined by electrophoresis. This technique involves causing DNA to migrate within a gel under the influence of an electric field. The negatively charged DNA fragments move towards the anode at a rate inversely proportional to their size. After electrophoresis the fragments are visualized by means of dyes, and their size is calculated. The restriction map can then be reconstituted

very practical and powerful: any fragment of DNA can be perfectly characterized by the analysis of the restriction products obtained by digestion with several enzymes.

1.2.2 Cloning and construction of libraries

Large molecules of DNA, especially chromosomes, are not directly accessible by experimentation, and must be cloned in order to become manipulable. Cloning consists of the insertion of a fragment of DNA into a vector, this vector being propagated in a cellular host. Through culture of this cell and purification of the vector, virtually limitless quantities of the cloned DNA fragment which one wishes to study may be produced. Bacterial cloning vectors are principally plasmids, phages and cosmids. These may carry fragments of which the maximum sizes are respectively about 10, 20 and 45 kb. More recently developed vectors are the BAC (Bacterial Artificial Chromosomes), P1 and PAC (P1-derived artificial chromosomes), wherein 100–300 kb fragments may be inserted.

Cloning may relate to an unique fragment of DNA, already purified. In this case one may proceed to sub-cloning. Alternatively one may seek to clone a complex population of DNA; this is the case in library construction, where the aim is to clone the most representative possible sample of the original population. A DNA library (Figure 1.5) comprises a group of clones representing the entire genome from which they derive; a complementary DNA library (cDNA library) is a collection of clones representing a group of messenger RNAs from a chosen tissue, previously recopied into DNA (cf. Chapter 5).

1.2.3 Hybridization techniques

Hybridization techniques allow the detection of a given sequence amongst a population of other sequences, with the aid of a probe, a fragment of DNA whose sequence is complementary to that which is sought. The probe is labelled by the incorporation of radioactive atoms or fluorescent groups. The detection of an homologous sequence is made possible by the specific matching of complementary sequences.

In the case of a Southern blot, the target DNA is a mixture of fragments, obtained by enzymatic digestion, separated by electrophoresis and transferred to a membrane (Figure 1.6). It is then possible to determine if a fragment corresponding to the probe being used is present in the target DNA and, if it is, the size of the corresponding fragment. Hybridization also makes possible the screening of a library. In this case, the target DNA is the assembly of clones of the library distributed on a

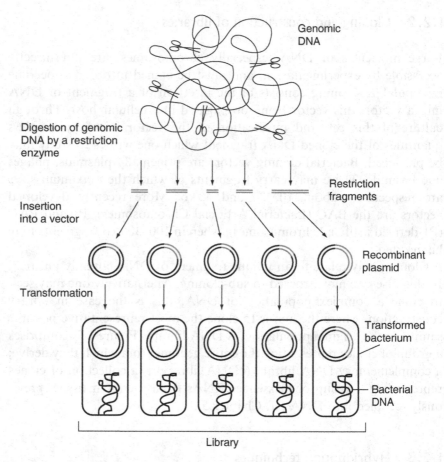

Figure 1.5 Construction of a genomic DNA library. The DNA is cut with restriction enzymes and inserted into a vector. The molecules obtained are introduced into a host cell (by transformation). The clones are individually cultured

membrane, which makes it possible to identify those clones containing a sequence corresponding to that of the probe.

Northern blots allow the detection of gene transcription (Figure 1.7). In this case, the fragments separated by electrophoresis and transferred to a membrane are messenger RNAs originating from different tissues or stages of development. The signal obtained with a probe corresponding to a given gene shows in which tissue the gene is expressed (the size of the messenger may also be obtained). Northern blots allow the detection of an mRNA represented once in every 100 000 mRNA molecules.

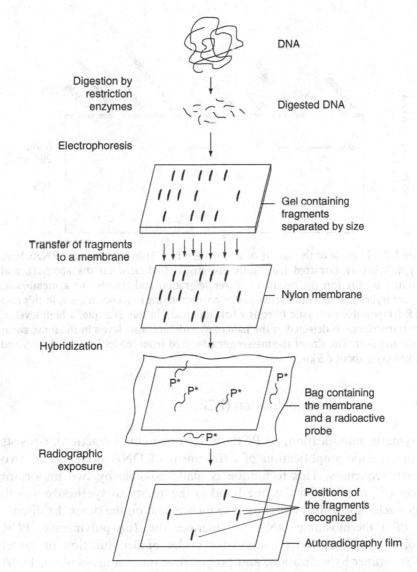

Figure 1.6 Southern blot. The DNA is digested by a restriction enzyme. The fragments are then separated by size using electrophoresis. After migration, the fragments are transferred onto a nylon membrane and fixed, conserving their separation by size. This membrane is then incubated with a radioactive probe, which will hybridize specifically to its complementary sequences. The size of these fragments is determined by exposing the hybridized membrane to autoradiographic film

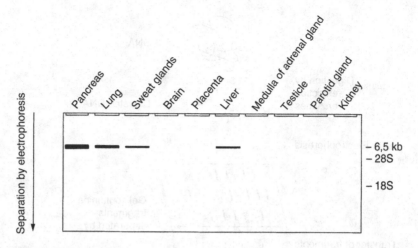

Figure 1.7 Example of the results of a Northern blot. Human messenger RNAs have been individually extracted from different tissues (indicated on the abscissa), and separated by size (on the ordinate). After migration and transfer to a membrane, they are hybridized to a radioactive probe corresponding to a known gene, in this case CFTR (responsible for cystic fibrosis when mutated). In this example, a high level of gene transcription is detected in the pancreas, with moderate levels in the lung, sweat glands and liver. The size of the messenger (deduced from the location of the 18S and 28S RNAs) is about 6.5 kb

1.2.4   Enzymatic amplification (PCR)

Enzymatic amplification, or PCR (polymerase chain reaction), consists of the specific amplification of a fragment of DNA bounded by two known sequences. This technique is made possible by two important pieces of progress. On the one hand is the ability to synthesize at will oligonucleotide primers of chosen sequence, and on the other the discovery of a thermostable DNA polymerase, the *Taq* polymerase. PCR involves the repetition of successive cycles of denaturation of target DNA, primer hybridization, and polymerization of a molecule of DNA identical to that situated between the primers. Progress through the stages of these cycles is achieved by manipulating the temperature of the reaction mixture, which necessitates the use of a thermostable polymerase (Figure 1.8).

The applications of PCR are extremely diverse. We will describe two of the main ones. The first is the production of a significant quantity of a given fragment of DNA, so as to facilitate the analysis of the fragment. This application tends actually to replace cloning (although it cannot

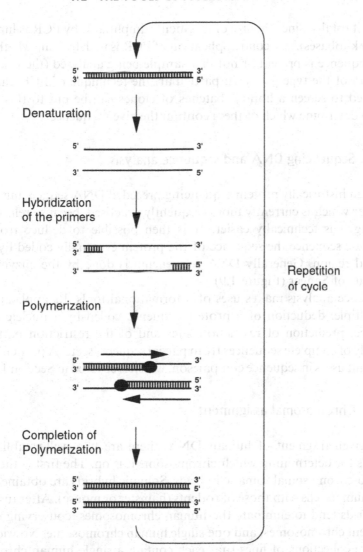

Figure 1.8 PCR technique. The initial reaction mixture includes the fragment to be amplified, a pair of oligonucleotide primers whose sequences are complementary to the ends of the fragment, *Taq* polymerase and free nucleotides. At 94°C, the two strands are separated (thermal denaturation), at 55°C the two primers each hybridize to the two ends, and at 72°C *Taq* polymerase (shown as a solid circle) synthesizes a new strand starting from the primer. Repetition of this cycle results in an exponential increase in the quantity of the initial DNA, each cycle producing a number of molecules equal to the amount present in the preceding cycle

replace it totally, since the size of fragments amplifiable by PCR is limited to a few kilobases). A second application of PCR is to determine whether a given sequence is present or not in a sample being analysed (the result is therefore of the type $+/-$). In particular, the technique of PCR can be employed to screen a library, batches of clones can be put to this test, so as to determine which of them contain the given sequence.

1.2.5   Sequencing DNA and sequence analysis

Although historically protein sequencing preceded DNA sequencing, it is the latter which is currently more frequently practised, because relatively speaking it is technically easier. It is then possible to deduce from a nucleotide sequence the sequence of the protein eventually coded by the analysed region. Generally DNA sequencing is done by the enzymatic technique of Sanger (Figure 1.9).

Sequence analysis makes uses of informational tools. Its applications are multiple: deduction of a protein sequence coded by a nucleic acid sequence, prediction of restriction sites and of the restriction pattern, assembly of complete sequences from partial sequences, etc. A particularly important use is in sequence comparison, which is covered in Section 1.3.3.

1.2.6   Chromosomal assignment

For a given fragment of human DNA, there are two well-established methods for determining which chromosome it is on. The first is the use of mono-chromosomal somatic hybrids. Somatic hybrids are obtained by fusing human cells with those of rodents (hamster or mouse). After fusion, the hybrids tend to eliminate the human chromosomes, conserving only the rodent chromosomes, and one single human chromosome. Nowadays there are collections of lines that each contain a single human chromosome. These hybrids are useful for determining (by hybridization or PCR) from which human chromosome a given sequence or clone has come.

A second method is *in situ* chromosomal hybridization, or FISH (fluorescent *in situ* hybridization). To determine the position of a fragment of DNA in the genome, the fragment is labelled and hybridized to a collection of chromosomes. Chromosome staining via Giemsa is used as a marker to determine the position of the signal. This stain allows the visualization of about 400 chromosomal bands, alternately dark (G bands, Giemsa) and light (R bands, reverse). This method of cyto-

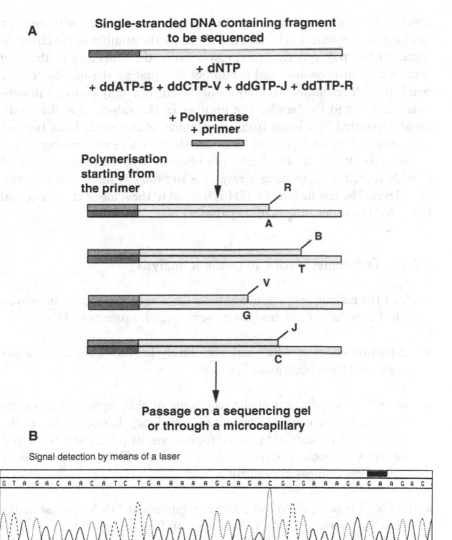

Figure 1.9   Sequencing. (A) the fragment of DNA which one wishes to sequence is cloned adjacent to a known sequence, and the hybrid is produced in single-stranded form. An oligonucleotide whose sequence is complementary to the known sequence is used as a primer to polymerize the complementary strand of the unknown sequence. The reaction is carried out in a tube containing four modified nucleotides (dideoxyribonucleotides, ddNTP), each labelled with a different-coloured fluorescent group: dideoxyadenosine (ddATP) with a blue fluorophore (B), dideoxycytosine (ddCTP) with a green fluorophore (G), dideoxyguanine (ddGTP) with a yellow fluorophore

genetic description offers a practical and reproducible technique for marking the genome. Each band is defined by the number of the chromosome, a letter p or q designating, respectively, the short arm or the long arm of the chromosome, and a number designating the number of the band. If high-resolution techniques are used, one may distinguish sub-bands interior to the bands. The number of the sub-band is then indicated, separated by a point from the number of the band. Thus 16p13.3 represents sub-band 3 of band 13 of the short arm of chromosome 16.

The resolution of localization obtained by mono-chromosomal hybrids is mediocre, because it reveals a site only at the whole-chromosome level. The resolution of FISH is limited to the scale of chromosomal bands (of the order of tens of megabases).

1.2.7   Techniques specific to genomic analysis

Some of the techniques presented above have been specifically improved so as to be useful for the analysis of very large fragments of DNA:

- restriction enzymes which cut very rarely (of the order of once per megabase) have been identified;

- the field of application of the technique of electrophoresis has been extended to fragments of great size (several megabases), thanks to the technique of pulsed field gel electrophoresis; its principle is to submit the DNA molecules not to a uniform electric field, but to fields whose orientation changes periodically;

- finally, it is possible to clone large fragments of DNA in yeast artificial chromosomes (YAC). The size of fragments clonable in this system (of the order of the megabase) greatly exceeds all the bacterial cloning systems (Figure 1.10). Nevertheless YACs present some prob-

---

(Y), and dideoxythymine (ddTTP) with a red fluorophore (R). The ddNTPs are diluted with dNTPs, in such a fashion that polymerization may be carried out over a certain length, but will halt once a ddNTP is added (absence of the OH on the 3' position of the ddNTP precludes the addition of subsequent nucleotides). (B) Once the polymerisation reaction is terminated, its results are separated by electrophoresis, and each type of molecule is detected thanks to its fluorescent marker, which is excited using a laser. The size of the molecules detected and their type of fluorescence then allows reconstruction of the complete sequence of the fragment

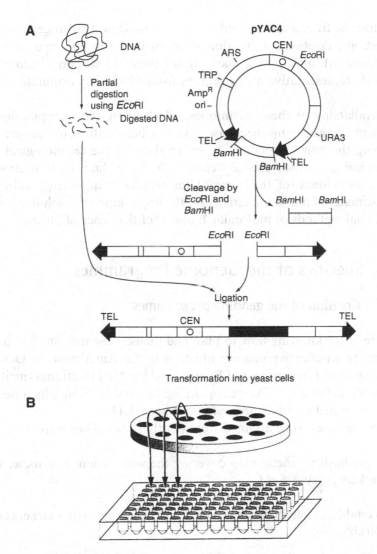

Figure 1.10 YAC cloning. (A) The vector comprises all the sequences necessary for maintenance and segregation of the YAC in yeast as if it were a real chromosome: a centromere (CEN), two telomeric extremities (TEL), and an origin of replication (ARS). Selective markers (URA3 and TRPI) and a cloning site (*Eco*RI) are also incorporated. The *Bam*HI–*Bam*HI fragment is first eliminated so as to liberate the telomeric extremities, then human DNA fragments are ligated into the *Eco*RI site. The constructs obtained are introduced into yeast cells. (B) Yeast clones are individually distributed into 96-well microplates arranged in 12 columns (1–12) and eight rows (a–h). Each well contains a single clone, completely identified by the number of its microplate and its coordinates thereon (e.g., there are 344 microplates in the CEPH / Généthon YAC library). It is then possible specifically to re-extract the recombinant DNA from every clone of interest

lems: the fragments inserted sometimes undergo rearrangements (insertions, deletions, formation of chimaeras between two or more fragments), and these clones cannot therefore be considered absolutely representative of the regions from which they originate.

The availability of these techniques, which allow the manipulation of fragments approaching megabase size, has been critical for progress in mapping the human genome. In effect they fill the technological and cytological gap which existed between the lower limit of resolution of genetic techniques (of the order of the centiMorgan, corresponding to approximately 1 mb in humans), and the upper limit of resolution of the traditional methods of molecular biology (of the order of 50 kb).

1.3　Specifics of the Genome Programmes

1.3.1　Creation of the genome programmes

The credit for knowing how to plan and finance research on the human genome in an effective manner must go to the Americans. In October 1990, the DOE (Department of Energy) and the NIH (National Institutes of Health) defined in a five-year plan the broad directions which needed to be integrated into this programme (Figure 1.11).

The objectives to be attained were clearly stated. They were:

- to establish a genetic map covering the whole human genome, with markers every 2–5 cM.

- to establish a physical map of the human genome, with markers every 100 kb or so;

- to improve the performance of automatic sequencing; complete sequencing of the human genome was considered to be an objective more long-term than the five-year interval covered by this first plan;

- to sequence the genome of several model species;

- to develop informational tools allowing the treatment, storage and communication of the data produced.

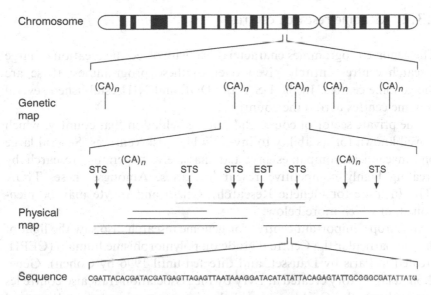

Figure 1.11 Objectives of the Human Genome Programme. Genetic mapping required the placement of regularly spaced markers throughout the genome, with intervals less than 5 cM. Physical mapping aimed to cover the genome by clones of genomic DNA, ordered by means of markers spaced every 100 kb. Transcriptional mapping has allowed the identification of 30 000 genes

In 1993 a new five-year plan was redefined, with the aim of refining the goals to be attained in the light of the objectives already accomplished. One new objective was the identification of the largest possible number of genes. The incorporation of this supplementary objective was due to the success gained by the programme of intensive cDNA sequencing, and to the necessity from the point of view of positional cloning to localize these cDNAs on the genome. The last plan was drawn up in 1999. It proposed the completion of the first stage of the human genome sequence, in the form of a working draft, an incomplete and unfinished sequence, but covering 90 per cent of the genome, each base having been read four or five times on average. The complete and finished sequence, with eight to 10 readings per base, was to be obtained in 2003 by this plan.

The American objectives were thus clearly defined and realistic, given the financial resources at their disposal. Less ambitious genome programmes were developed in other countries, principally the UK, Japan, France, Germany and China.

1.3.2   Genome research centres

The genome programmes characteristically involved the creation of large research centres entirely given over to these programmes: these are the genome centres. In the USA, the DOE and NIH established several genome centres across the country.

The private sector of course did not lag behind in that country, which is well known for its ability to invest in high-risk ventures. Several large pharmaceutical companies invested massively in genome research by creating highly competitive research centres. Amongst these, TIGR (The Institute for Genetic Research), Celera and Incyte may be mentioned, of which more below.

In Europe, important centres for genome research also saw the light of day, in particular the Centre d'étude du polymorphisme humain (CEPH, created in Paris by Dausset, and directed until 1996 by Cohen), Généthon (a laboratory created at Évry by the Association française contre les myopathies), the Centre national de séquençage (located at Évry and directed by Weissenbach), and the Sanger Centre (established at Hinxton by the Wellcome Trust, and directed by Sulston).

1.3.3   The growing importance of bioinformatics

Bioinformatics plays an essential role in genome research. This is because of the complexity and quantity of the information generated by these researches, which increases in an extremely rapid fashion (mapping or sequencing data). Informatics is involved in automatic data acquisition, and in the archiving, distribution and exploitation of the data. Most of the large genome research centres have a server putting their local database at the disposal of the scientific community.

There also exist many international databases, each prioritizing a certain type of information: sequences (GenBank, Swissprot, Golden-Path, ENSEMBL), mapping (NCBI—National Centre for Biotechnology Information), genetic diseases (OMIM—On-line Mendelian Inheritance in Man), etc. The growth of these databases is almost exponential. About 20 000 organisms are represented in these databases: 10 per cent of sequences are of viral origin, 20 per cent come from eubacteria, 1 per cent from archaebacteria and 69 per cent from eukaryotes. In this last group mammalian sequences are most frequently represented, and 80 per cent of these are human.

Access to these databases is achieved via the Internet. Because of the enormous mass of information contained in them, and the rapidity with which they evolve, the ability to access them quickly, efficiently and frequently is a necessity for all genetic laboratories.

One particularly interesting application is the comparison of protein sequences, which allows the detection of similarity with an already-known sequence. This type of analysis is widely practised for all newly acquired cDNA sequences which – after informatic translation – are compared with the collection of sequences already in the databases. Three situations may thus become apparent: (i) the sequence has no similarity to any of those already collected; (ii) it is identical to an already-known sequence; and (iii) it is similar, but not identical, to an already-known sequence (Figure 1.12). One might then be looking at a homologous sequence from another species, sequences of the same multi-gene family or of the same superfamily (in the latter case the degree of similarity is weaker).

This comparison allows a certain measure of prediction of protein function: if a new sequence shows high similarity to that of an already-known protein of exprimentally established function, it is probable that the new protein's function is similar to that with which it has been aligned. Equally, the same protein activity is often carried out by proteins of similar sequence.

This type of prediction is, however, not 100 per cent reliable. The bacterial parasites *Listeria monocytogenes* and *Shigella flexneri* both

```
H. sapiens       LYPSIMMAHNLCYTTLLRPGTAQKL    GLTE DQFIRTPT GDEFVKTSVRKGLLPQILENLLSARKRAKAELAKETDPLRRQ
B. taurus        --------------------A----    ---- ----K--- ------A-----------------------------------
M. musculus      --------------------A----    --KP -E--K--- ------S------------------------Q--------
D. melanogaster  ----------------VLG--RE-LRQQEN-QD --VE---A NNY---SE--R----E---S--A------ND-KV----FK-K
C. melanogaster  ------I----------         -S PQ-VEN EDY----S -QY-ATK-K-R----E---DI-A------NDMKN-K-EFK-M
O. sativa        --------Y----C--VP-ED-R-- N-PP ESVNK--S -ET---PD-Q--I--E---E--A--------D-KEAK--FE-A
B. subtilis      ---------------CNKA-VER-    --KIDEDYVI--N    Y--T-KR-R-I---I--DE---------KD-RD-K--FK-D
P. abyssi        -----IIT--VSPD--    N-  E-CKN   YDIA-QV-HK-C- DI P-FI-SL-GH--EE-QKI-TKMKETQ--IEKI
V. cholerae      -----IRSFLIDPLG-I    EG-KLPI-KQA -HAVPGFR -GQ-HR-    -HF--EMI-K-WA--DE--RNQE-AFSQAIKI
```

Figure 1.12 Comparison of protein sequences. The protein sequences of a region of DNA polymerase δ from eight species are compared with that of humans. The conserved sequences are identifed by hyphens; some regions are only present in one or several species but not in others, represented by gaps when they are absent. The degree of conservation of these sequences in comparison to that of man is inversely proportional to the evolutionary distance which separates them (*P. abyssi, Pyrococcus abyssi; V. cholerae, Vibrio cholerae*)

induce actin during infection, mediated by protein ActA in the first case and IcsA in the second, but these proteins are not at all similar.

Informatic analysis of genomic sequences has also assumed considerable importance, because of the increasing power of sequencing complete genomes. One difficulty of genomic sequence interpretation is the identification of exons which are, especially in humans, drowned in the midst of vast non-coding regions (introns or intergenic regions). It is sometimes possible to identify exons by comparing the genomic sequence with that of the collection of cDNAs obtained from the same species: genomic regions which align 100 per cent with messenger sequences almost certainly represent exons. Another approach is by comparison between the genomes of different species: detection of a similarity between the genome of one species and that of another is most often significant. In effect, only sequences that are functionally constrained are likely to be conserved in evolution. This is particularly the case in the majority of coding sequences, which diverge much less quickly than non-coding sequences (Figure 1.13). Other techniques are discussed in Section 4.1.4.

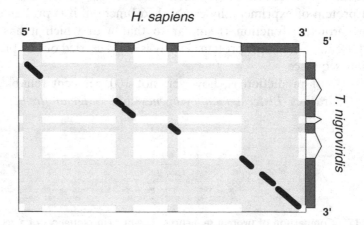

Figure 1.13 Comparison of genomic sequences. This figure shows a comparison between the genomic sequences obtained for *T. nigroviridis* (on the abscissa) and humans (on the ordinate), each of which contains the gene encoding the c-fos protein, whose exons are shown as grey rectangles, joined by broken lines representing the introns. Within the rectangle which separates the two genes, several diagonals show the regions conserved between the two genomic regions. These conservations are only found for regions corresponding to exons (grey patches), with none found for regions corresponding to introns (white patches)

1.3.4 Automation and robotics

Another particular feature of the Genome Project is the repetitive character of a number of tasks, which has led to the development of robots able to replace humans in their performance, e.g. the Genomatron of the Whitehead Institute/Massachusetts Institute of Technology (WI/MIT) simultaneously carries out and analyses 150 000 PCRs; Généthon has constructed 20 machines capable of automatically carrying out 16 Southern blots at once.

One extremely important breakthrough was the automation of sequencing. Automatic sequencers allow the automatic performance of the stages of sequence separation and acquisition. The highest-performance sequencers of today are capable of analysing 500 bp of each of 64 samples in 4 h, or 1000 bp of each of 32 samples in 10 h. However, a new generation of sequencers appeared in 1998; these are multicapillary sequencers, which allow the sequencing of 96 samples in 4 h.

1.4 The Species Analysed

1.4.1 The main branches of life

Prokaryotes are distinguished very clearly from eukaryotes by the absence of a nucleus and organelles, and the size of their genome (Table 1.1). The majority are unicellular, but some form multicellular organisms. Their size is much smaller than that of eukaryotes, of the order of $1-5\,\mu m$, and their cell wall plays an important protective role.

Table 1.1 Comparison of prokaryotes and eukaryotes

	Prokaryotes	Eukaryotes
Nuclear membrane/nucleus	Absent	Present
Mitosis/meiosis	Never occurs	Occurs
Mitochondria (chloroplast), Golgi apparatus, endoplasmic reticulum	Absent	Present
Ribosomes	Small size (70S = 50S + 30S)	Large size (80S = 60S + 40S)
Size	$1\,\mu m$	$10\text{-}30\,\mu m$
Chromosome number	Generally only one	Several
Introns	Absent or very rare	Often present

Initially, the primary classification factors used for prokaryotes did not allow the establishment of a very reliable phylogenetic tree. In the end it was molecular biology which showed itself to be most useful, allowing the distinction within the prokaryotes of the archaebacteria and eubacteria (a division proposed by Woese in 1977). Archaebacteria and eubacteria differ particularly in their cytoplasmic membrane composition, in their RNA polymerases, and their ribosomes.

Eukaryotes are characterized by the presence of a nucleus (bounded by a double nuclear membrane), within which mRNAs are modified (in particular by splicing, addition of a 5' cap, and 3' polyadenylation), the presence of organelles (mitochondria, and chloroplasts in plants), and the existence of separate chromosomes (the number of which is invariable for a given species), within which the DNA is associated with histones.

1.4.2 Prokaryotes

In the biosphere, prokaryotes represent an essential collection of biodiversity. Their capacities for adaptation can achieve extraordinary levels: eubacteria have been recovered alive from the 10 000-year-old remains of mastodons, and certain archaebacteria are adapted to survive at low or very high temperatures (psychrophiles or thermophiles), or in sites very rich in salts (halophiles) or with very acidic pH (acidophiles).

The majority of prokaryotes live autonomously, but many eubacteria interact with other species. These relationships may be symbiotic or parasitic. Symbiosis has certainly played a very important role in the evolution of life: it seems to be the cause of the presence of organelles in eukaryotes (mitochondria and chloroplasts). Parasitism correlates with certain diseases, for which these organisms are particularly responsible in humans (no archaebacterium has yet been identifed as pathogenic).

Genome size varies amongst these species between less than 600 kb and several megabases (Table 1.2). The chromosomes are generally circular, but there are also examples of linear chromosomes (for example in *Borrelia burgdorferi*, *Rhodococcus fascians* and *Streptomyces coelicolor*). The ends of linear bacterial chromosomes, like those of eukaryotes, are stabilized by telomeres. Curiously, given that in many respects archaebacteria seem to be closer to eukaryotes than eubacteria, no linear chromosome has yet been discovered in this group.

The genome is generally represented by a single chromosome, but several chromosomes are present in certain species, e.g. *Rhodobacter*

Table 1.2   Genome sizes of various prokaryotes

Archaebacteria	
Haloferax volcanii	4.1 Mb
Sulfobolus acidocaldarius	2.8 Mb
Halobacterium halobium	2.4 Mb
Thermococcus celer	1.9 Mb
Pyrococcus horikoshii	1.7 Mb
Eubacteria	
Myxococcus xanthus	9.4 Mb
Anabaena sp.	6.4 Mb
Shigella flexneri	4.6 Mb
Staphylococcus aureus	2.8 Mb
Thermus thermophilus	1.7 Mb
Ureaplasma urealyticum	0.75 Mb
Mycoplasma genitalium	0.6 Mb

sphaeroides, Brucella melitensis, Bacillus cereus, Deinococcus radiodurans and Vibrio cholerae. There are even examples where a linear chromosome coexists with a circular one (*Agrobacterium tumefaciens*).

Also present may be extrachromosomal structures, in the form of independently replicating DNA molecules: plasmids. Their size can vary greatly, from 1000 nucleotides up to more than 1 Mb. They are often circular, but may be linear. The genes carried by plasmids are very diverse. They are not essential for the organism if it is in a favourable environment, but under certain conditions they can confer a selective advantage. Some of them encode proteins that confer resistance to antibiotics or heavy metals.

For many species, the distinction between chromosome and plasmid is clear. On the one hand the size of the latter is often small, on the other hand they are not generally indispensible. However this distinction is less evident for the megaplasmids (plasmids of great size), which often contain genes coding proteins indispensible for life.

The analysis of the genome of *Escherichia coli* has been undertaken for several reasons: this bacterium is present in humans, it is the favourite species of researchers in the field of molecular biology, and the variety 0157:H7 is a pathogen. The genome of *Bacillus subtilis* has been analysed because of the widespread use of this species in the agro-alimentary field (production of proteases and amylases). In Japan, for example, the 'natto' strain is used to prepare fermented dishes based on soya.

1.4.3   Eukaryotes

The actual number of species of eukaryotes is difficult to evaluate; proposed figures vary between 5 and 50 million. Many are multicellular organisms – fungi, higher plants, metazoans – amongst which are humans.

Among the diverse species of eukaryotes there exists an extraordinary variation of genome size, but this does not correlate with organismal complexity (Table 1.3). Thus, amongst vertebrates, the genome of *Protopterus aethiopicus* (fish, 140 000 Mb) is more than 43 times larger than that of humans (3000 Mb), and the latter is eight times larger than that of

Table 1.3   Number of chromosome pairs and genome sizes of various eukaryotes

	Number of chromosome pairs	Genome size
Protozoa		
Plasmodium falciparum	14	30 Mb
Leishmania major	36	35 Mb
Schistosoma mansoni	8	270 Mb
Fungi		
Schizosaccharomyces pombe	3	14 Mb
Saccharomyces cerevisiae	16	13 Mb
Invertebrates		
Dictyostelium discoideum	6	34 Mb
Caenorhabditis elegans	6	100 Mb
Drosophila melanogaster	4	180 Mb
Vertebrates		
Tetraodon nigroviridis	21	400 Mb
Gallus gallus	39	1200 Mb
Danio rerio	25	1700 Mb
Homo sapiens	23	3000 Mb
Rana esculenta	13	8400 Mb
Necturus maculosus	19	50 000 Mb
Protopterus aethiopicus	17	140 000 Mb
Metaphytes		
Arabidopsis thaliana	5	125 Mb
Oryza sativa	12	420 Mb
Zea mays	10	5000 Mb
Lilium longiflorum	12	90 000 Mb
Fritillaria assyriaca	12	125 000 Mb

Tetraodon nigroviridis (400 Mb). Amongst the angiosperms, the genome of *A. thaliana* covers 125 Mb, whereas that of *Fritillaria assyriaca* is 130 000 Mb. These differences in size do not reflect gene content; they are largely due to the presence of repeated sequences.

Chromosome number per genome is also very variable: the most extreme figures have been found in the ant, *Myrmecia pilosula* (a single pair of chromosomes), and in the fern, *Ophioglossum reticulatum* (630 pairs). Even between very closely related species, chromosome number can differ greatly: of two species of deer which are very similar phenotypically, one (*Muntiacus reevesi*) has 46 chromosomes and the other (*Muntiacus muntjac*) six (in females) or seven (in males). In humans, the identification of 23 pairs (22 autosomes and one pair of sex chromosomes) was made by Tijo and Levan in 1956.

The imposing complexity of the human genome encouraged an initial bias towards model organisms for the development of a certain number of genetic studies. This was justified, on the one hand, by the structural and functional uniqueness of life and, on the other, the lesser genome size of these species (Table 1.4). These model species have several interesting characteristics from the experimental point of view: short life cycle, ease of growth or culture and numerous offspring (Table 1.5).

Just as with *E. coli* or *B. subtilis*, the yeast *Saccharomyces cerevisiae* has been one of the work-horses of geneticists, cellular and molecular

Table 1.4 Comparison of the characteristics of the genomes of model organisms and humans

Organism	Systematic position	Haploid chromosome number	Genome size (Mb)	Number of genes
Escherichia coli	Bacteria	Unique circular	4.7	4300
Saccharomyces cerevisiae	Ascomycete	16	13	6200
Schizosaccharomyces pombe	Ascomycete	3	14	4900
Caenorhabditis elegans	Nematode	6	100	19 100
Arabidopsis thaliana	Angiosperm	5	125	25 000
Drosophila melanogaster	Insect	4	180	13 400
Mus musculus	Mammal	20	2800	30 000[a]
Homo sapiens	Mammal	23	3200	30 000[a]

[a]Estimate.

Table 1.5 Characteristics of model eukaryotic organisms. In *C. elegans*. about 280 individuals are obtained from the hermaphrodite, 1000 by male-hermaphrodite crossing

Species	Size	Generation time	Descendants per generation
Saccharomyces cerevisiae	5 μm	1 h 30	2 (mitosis)/4 (meiosis)
Arabidopsis thaliana	30 cm	10 weeks	Several thousand
Caenorhabditis elegans	1 mm	3–5 days	280/1000
Drosophila melanogaster	2 mm	10 days	Several hundred
Mus musculus	10–15 cm	2.5 months	4–12

biologists for many years. This unicellular fungus can maintain itself equally well in the haploid and diploid states, and can live without mitochondria. It is the first eukaryote for which the whole genome has been sequenced. More recently (2002), the genome of *S. pombe* has also been completely sequenced. The studies undertaken on this species in particular have aided the elucidation of control of cell division, mitosis and meiosis, and DNA repair.

The crucifer *Arabidopsis thaliana* is the model organism of reference for biologists working on higher plants. The main interest of this plant is its small genome size, the smallest known amongst the angiosperms (125 Mb), a peculiarity which has led some authors to nickname *A. thaliana* 'the green drosophila'. Its genetics are simple, auto-fertilization or outcrossing being equally possible.

The nematode *Caenorhabditis elegans* is an oligocellular model organism greatly valued by embryologists and neurobiologists. Hermaphrodites have two sex chromosomes, XX, and males only one (there are no females). This animal has the advantage of being completely transparent, which allows one to follow the fate of each cell throughout its development. Thus it has been possible to determine the exact ancestry of each cell of this organism from fertilization up to adulthood (959 cells). Equally exactly understood is the assembly of the synaptic junctions established by each of the 302 neurons. From this point of view, no other multicellular organism is so well understood. It was the first eukaryote for which the whole genome was sequenced (in 1998).

The fly *Drosophila melanogaster* is also a well-studied model organism, ever since the pioneering genetic studies on this animal carried out by Morgan at the beginning of the twentieth century. Its genome comprises four pairs of chromosomes, females being XX and males XY. A particularly interesting characteristic of this dipteran is the existence of polytene chromosomes in certain of its tissues. These chromosomes comprise about 1000 identical molecules of DNA, which are the product of multiple DNA replications without strand separation. The strands appear parallel and united, which gives them an easily visible structure that is very useful in physical and genetic mapping. One can thus recognize about 5100 bands by simple staining and microscopic observation of these chromosomes, and the resolution obtained is of the order of some tens of kilobases (the average band size is 26 kb).

The size of the mouse genome (*Mus musculus*), at 2800 Mb, is comparable to that of humans (3000 Mb). Nonetheless, as a mammal, this is obviously the closest we have to a model organism. The possibility of performing any cross which is desired and the existence of pure lines and of a large number of identified mutations (some of which resemble human genetic diseases) make the mouse an irreplaceable asset for genetic studies on mammals.

2 Linkage Maps

Because of the great complexity of genomes, progress in studying them often depends upon the establishment of maps, which allow orientation along chromosomes by the aid of markers. These markers are fragments of DNA isolated and cloned from the appropriate genome, and their order upon it then needs to be established. There are essentially two types of map, which differ according to the type of fragments to be ordered, and according to the units used to measure the distances separating the fragments: linkage maps (genetic maps and irradiation maps) and physical maps (whose study is the subject of the next chapter).

Linkage maps are constructed from markers which correspond most often to short fragments of genomic DNA, each defining a unique locus. These maps indicate the positions of the markers relative to each other. Once established, they serve as a reference for localizing any other marker or gene in the genome.

For a genetic map, the markers are ordered through the analysis of their segregation over the course of generations. Until recently genetic maps were the main type of linkage map. However, a new type of linkage map has subsequently appeared: these are maps constructed using radiation hybrids, which we will discuss at the end of this chapter.

2.1 Tools and Methods in Genetic Mapping

2.1.1 The genetic maps of *Drosophila* and humans

The idea of the genetic map goes back to 1913, with the work of Morgan and Sturtevant on *Drosophila* (these studies contributed to the elaboration of the chromosomal theory of heredity). The conclusions of their

Genome, Transcriptome and Proteome Analysis   by Alain Bernot
© 2004 John Wiley & Sons, Ltd   ISBN 0 470 84954 1 (HB)   ISBN 0 470 84955 X (pbk)

work showed that two markers localized on the same chromosome are generally transmitted together to the next generation. However, during meiosis, markers situated on the same chromosome may become separated if there is a crossing-over event in the region between them (Figure 2.1). The probability of such an event is proportional to the distance between the markers. The frequency of recombination thus

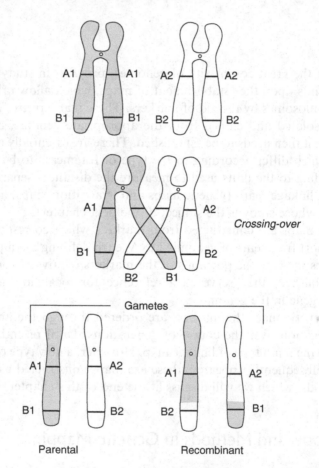

Figure 2.1 Establishment of crossing-over during meiosis. (A) Before meoisis, the two homologous chromosomes are each composed of two sister chromatids. Each chromatid carries respectively at the loci A and B the alleles A1 and B1, A2 and B2. (B) During meiosis, chromatids from each homologous chromosome may exchange genetic material between the loci A and B. (C) The resulting gametes consequently exhibit mainly the combinations A1 with B1 and A2 with B2, but a fraction carry the alleles A1 with B2 or A2 with B1. This fraction is larger the further apart locus A is from locus B (but may not exceed 50 per cent)

reflects the distances between markers, and the unit of distance is the centimorgan (cM). A centimorgan corresponds to a recombination frequency of 1 per cent (one crossing-over per hundred meioses).

From these pioneering studies sprang two critical observations concerning the establishment of genetic maps. The first is that one cannot establish distances between markers unless they are polymorphic, that is to say unless they are represented by different and identifiable alleles in the population. If this single condition is satisfied, then one can follow the segregation of markers over the course of generations. At the level of DNA, this polymorphism reflects variations between the sequences of the different alleles at each of the corresponding loci, which are eventually translated into observable phenotypic modifications. The second observation is that the distance between two markers is calculated by means of statistical analysis. These calculations must therefore be carried out on populations large enough to give statistical significance.

Genetic mapping was regularly advanced using numerous animal and plant species. The markers utilized were initially mutations that caused phenotypic variations, protein polymorphisms, or cloned genes. Unfortunately, the establishment of a human genetic map lagged well behind: on the one hand, the markers used in model animals are either rare or insufficiently polymorphic in humans; and on the other, human families are generally small in size.

2.1.2   Tools for genetic mapping in humans

Obtaining a genetic map composed of markers covering the whole human genome was therefore an objective of the first importance: such a map would allow the identification of the region and the gene responsible for a given genetic disease, or improve the performance of prenatal or pre-symptomatic diagnoses. A genetic map would also allow the establishment of anchorage points for the physical map of human chromosomes.

It is the progress of molecular biology since the 1970s that has allowed the development of collections of markers useful for human genetics, in the form of fragments of genomic DNA each defining an unique locus in the genome. These fragments have been chosen for their polymorphism, which manifests itself by subtle – but detectable – sequence differences between alleles. Two types of markers have been successively developed: RFLP (restriction fragment length polymorphism) and microsatellites.

Most of these DNA fragments do not correspond to translated sequences, and their variations are not manifest at the phenotypic level by the modification of a directly observable character: these are anonymous markers.

The second indispensible element for the establishment of a genetic map is the availability of DNA coming from large families, which allows the study of marker segregation over the course of generations. Dausset deserves the credit for having selected a group of families of great genetic value (containing a high number both of generations and of numbers of children), and above all for having put the collection at the disposal of the scientific community, through the CEPH. Thus, all the data produced by the ensemble of laboratories working on the DNA of these families is cumulative.

2.1.3   RFLP markers

An RFLP occurs when a fragment of cloned genomic DNA being used to probe a restriction enzyme digest of genomic DNA detects fragments whose sizes differ from one individual to another. This polymorphism can be the result of restriction site distribution, or the presence of differently repeated minisatellites. The size of the fragments detected depends on the positions of the cut-sites which surround or are internal to the region recognised by the probe, or on the number of repetitions of the minisatellite. RFLPs are chosen so as to recognize a unique locus, and the different alleles are represented by the possible cuts in this locus (Figure 2.2).

In practice, an RFLP analysis is done by using a DNA fragment as a probe of a Southern blot comprising genomic DNA from different individuals digested by a restriction enzyme. Each cutting pattern obtained is an allele of the marker under consideration. Each allele depends only on the sequence of the genomic fragment under consideration, and segregates in a Mendelian fashion over the course of generations. Thus it is possible to establish genetic distances between several such markers, or between a marker and a given characteristic, or to construct a map of the human genome.

RFLP analysis led to the localization of the genes responsible for common genetic diseases, such as Duchenne muscular dystrophy (1982), Huntington's chorea (1983) or cystic fibrosis (1985). This type of research led in 1986 to the first positional clonings of genes responsible for genetic

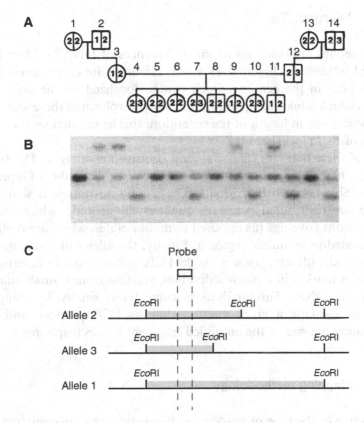

Figure 2.2 Genotyping by RFLP. (A) Genealogy of the family studied (squares represent males, circles females). (B) DNA from the individuals is digested with a restriction enzyme, the fragments are separated on a gel, transferred to a membrane and hybridized to a labelled probe. The size of the fragments revealed is specific for each allele: grandmother (1) has two copies of allele 2; grandfather (2) has alleles 1 and 2; grandmother (13) has two copies of allele 2; grandfather (14) has alleles 2 and 3. Children (4), (7) and (10) have the alleles 2 and 3, children (5), (6) and (8) have two copies of allele 2, and children (9) and (11) have alleles 1 and 2. (C) Interpretation of the RFLP, here detected by the enzyme EcoRI. The sizes of the revealed fragments depends on the presence or absence of restriction sites. Hatching represents the size of the fragment recognized by the probe

diseases, in particular that responsible for Duchenne muscular dystrophy. These mapping exercises were restricted to small fractions of the genome, but they clearly showed that RFLP mapping was an effective method for localizing human genes in general. This type of marker was thus used to establish the first human genetic map.

2.1.4   Microsatellites

Microsatellites are made up of short sequences of DNA (1–13 bp long) repeated in tandem (Figure 1.3). Each microsatellite corresponds to an unique locus in the genome which can be localized by the unique sequences which flank the repetition. The polymorphism of these loci is due to the variations in length of the repetition: this length defines the alleles of each of these loci (Figure 2.3).

Use of these markers carries several decisive advantages. The first is their very high degree of polymorphism, because the number of repeats is very variable between individuals. The second advantage is that these markers are distributed evenly throughout the genome, which ensures homogeneous coverage (as opposed to minisatellites, which are preferentially located in telomeric regions). Finally, the alleles of these markers are easily identifiable, since a single PCR suffices to characterize the alleles of a marker in a given individual, starting from a small quantity of genomic DNA. This analysis is carried out simply by using the sequences flanking a given microsatellite as PCR primers, and then determining the size of the amplified fragment by electrophoresis.

2.1.5   Mapping methodology

Irrespective of the type of marker used, genetic maps are constructed in the following manner:

- the markers are first of all assigned to chromosomes through the use of mono-chromosomal somatic hybrids, either by FISH (in the case of RFLP) or by PCR (microsatellites);

- for each marker, the alleles present amongst the individuals from large families of the CEPH are identified; such a characterization corresponds to the genotyping of the marker;

- initial analysis evaluates statistically the degree of linkage for all pairwise combinations of markers ('pairwise' informatic calculation). This first analysis allows the assignment of markers to linkage groups;

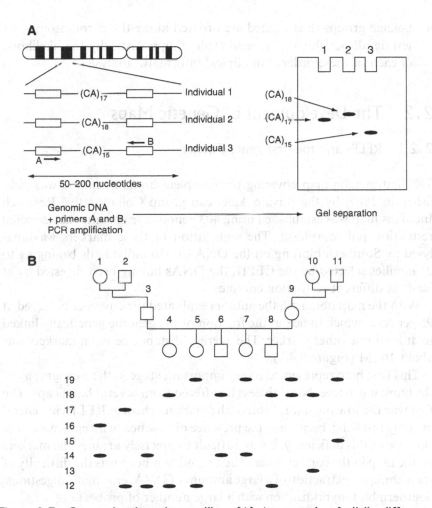

Figure 2.3 Genotyping by microsatellites. (A) An example of allelic difference between microsatellite sequences. The CA dinucleotide is repeated 17 times in individual 1, 18 times in individual 2, and 15 times in individual 3. The sequences flanking the repeats are identical between individuals, allowing the choice of primers for PCR (primers A and B). (B) Genotyping a microsatellite within a family. Primers corresponding to sequences flanking the repeat allow amplification of the microsatellite in different members of a family. Their size is determined by electrophoresis. The microsatellite is represented by two alleles of 16 and 14 dinucleotides in grandmother (1), 17 and 12 dinucleotides in grandfather (2), 18 and 13 in grandmother (10), 19 and 15 in grandfather (11). The father (3) has inherited the 12 and 14 repeat alleles, the mother (9) the 19 and 18 repeat alleles. Children (4–8) show respectively the allele combinations 18 and 12, 19 and 12, 18 and 14, 19 and 14, and 18 and 12

- linkage groups thus created are ordered along the chromosomes, by
  testing all possible orders, and evaluating statistically the likelihood
  of each of these orders ('multiplex' informatic analysis).

2.2   The Development of Genetic Maps

2.2.1   RFLPs and the first genetic maps

The first genetic map covering the complete human genome was published in 1987, by the private American group Collaborative Research
Inc. This map was established using 403 genomic fragments that detected
restriction polymorphism. The segregation of these markers was analysed by Southern blotting on the DNA of 310 individuals belonging to
21 families selected by the CEPH, the DNAs having been digested by at
least six different restriction enzymes.

With the map obtained, the authors evaluated the coverage achieved at
95 per cent, which indicates the fraction of the genome genetically linked
to at least one other marker. The average distance between markers was
about 10 cM (Figure 2.4A).

This first map represented a very important stage in the cartography of
the human genome. Nevertheless it suffered from several handicaps. The
first was the low polymorphism of the markers chosen: RFLPs in general
are only bi-allelic, because of the presence or absence of an enzymatic site.
Because of this deficiency, it was difficult to precisely arrange the markers
on the map in the correct order. The second handicap was the difficulty of
the technique (extraction of a large amount of DNA, enzymatic digestions,
Southern blot, hybridization with a large number of probes).

A second version of this map, with 1676 markers, was published in
1992 by a consortium associating the NIH with the CEPH (Figure 2.4B).
This version too was largely based on RFLPs (1317). It also included
other types of markers, in particular 339 microsatellites, the massive use
of which was to revolutionize human genetic mapping, sweeping away
the deficiencies mentioned above.

2.2.2   Microsatellites and modern maps

The development of microsatellite markers, extremely polymorphic and
simple to use, allowed the establishment of 'second generation' maps.

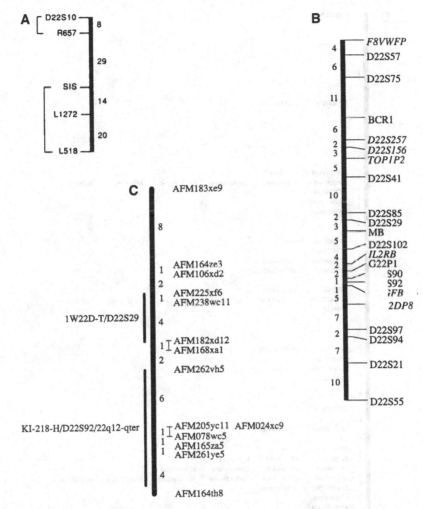

Figure 2.4 Comparison of successively obtained genetic maps. We have chosen chromosome 22 (the smallest) to illustrate progress in the resolution of genetic maps, but the resolution obtained for other chromosomes is similar. To the left of the chromosome, the markers are represented by a vertical trace, to the right is shown the genetic distances between them. (A) CRI map of 1987, with five RFLP markers mapped. (B) NIH/CEPH map of 1992 (19 RFLPs and four microsatellites). (C) First version of the Généthon map (14 microsatellites); two of these markers are so close that they cannot be

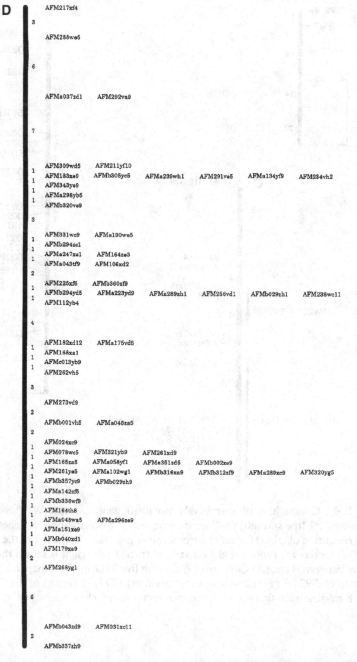

Figure 2.4 (*Continued*) distinguished genetically in the CEPH families (no recombination observed between AFM205ycll and AFM024xc9). (D) Final version of the Généthon map (67 microsatellites)

This option was pursued to the exclusion of all others at Généthon by Weissenbach, in a project which culminated in 1992 with the first human genetic map based upon this type of marker.

The Généthon map was established using $(CA)_n$ microsatellites. The CA repeat was chosen because of its abundance in the genome (there are 50 000–100 000 such repeats, uniformly distributed throughout the genome), and high degree of polymorphism. It was first necessary to isolate and sequence a large number of human genomic DNA fragments containing this type of repeat. Microsatellites with over 12 repeats, and whose degree of polymorphism and number of existing alleles were the highest were selected.

These markers were genotyped using a group of individuals from eight large CEPH families. Thus, 814 microsatellites could be mapped. These markers were separated on average by 4.4 cM (Figure 2.4C). Although less marker-rich than the CEPH/NIH map (814 markers as opposed to 1676), the Généthon map was nonetheless immediately imposed as a reference. There were three reasons for this: (i) microsatellites are much easier to use than RFLPs; (ii) the information relative to such markers is entirely composed of the sequences of the primers and their location, it can thus easily be archived and communicated in an informatic form; (iii) most importantly the degree of polymorphism of microsatellites is remarkably high.

2.2.3 Progress of the human genetic map

Other genome centres produced markers using non-CA-repeat microsatellite types. The Universties of Iowa and Utah also isolated 5150 markers based on di-, tri- and tetra-nucleotide repeats. Because all these markers were genotyped using the same reference DNA collection, it was possible to combine all of the data.

Finally, a second version of the Généthon map succeeded the first in 1994 (2066 microsatellites), followed by a final version in 1996 (5264 microsatellites) (Figure 2.4D). In this map the average interval size defined by these markers is 1.6 cM, which exceeds any resolution heretofore attained (Table 2.1).

2.2.4 SNP maps

A new type of marker is currently used during the construction of human genetic maps: this is the SNP (single-nucleotide polymorphism). The polymorphism of these markers is much reduced, represented by two

Table 2.1 Characteristics of the principal human genetic maps successively established since 1987. Several markers are so close that they rarely recombine: they are therefore mapped to the same genetic locus. This explains why the product of the number of markers and the resolution can give very different numbers

Year	Team	Marker type	Number	Resolution (cM)
1987	CRI	RFLP	403	10
1992	Généthon	Microsatellites $(CA)_n$	814	4.4
1992	NIH/CEPH	RFLP (and microsatellites)	1676	3
1994	Généthon	Microsatellites $(CA)_n$	2066	2.9
1996	Généthon	Microsatellites $(CA)_n$	5264	1.6

different bases located in the same genomic region, each of these sequences constituting a distinct allele (Figure 2.5). On the other hand, they are very abundant in our genome.

These markers have principally been obtained thanks to genomic sequencing of the two homologous chromosomes of the same individual, or of the same genomic region from different people. More than 2 million SNPs have been obtained, separated by about 1 kb. Mapping of SNPs in our genome was essentially achieved by comparison of their sequence with the already-existing human genome sequence data.

How are these alleles identified? The markers present in a given individual are first amplified by PCR, using primers flanking the SNP. Alleles may now be identified by hybridizing the amplified sequences to chips carrying oligonucleotides complementary to the different possible sequences. They can also be identified by mass spectrometry (cf. section 6.4.3), after polymerization with a primer complementary to the amplified sequence, which hybridizes immediately upstream of the SNP (Figure 2.5).

Figure 2.5 SNP markers. (A) Example of two distinct alleles representing an SNP, distinguished by an adenine (A) in the centre of the first allele, and a guanosine (G) in the second. (B) Identification of these alleles by PCR. The alleles are amplified by PCR and denatured. A probe is hybridized to the allele, immediately upstream of the SNP, and a polymerization is performed, but the only nucleotides supplied are dideoxynucleotides: thus, only one can be incorporated (absence of the 3′ OH precludes addition of subsequent nucleotides). (C) The products of this polymerization are subsequently analysed by mass spectrometry, which allows determination of the mass of the fragments obtained, and the differences between the weights of ddA, ddC, ddG and ddT are sufficient to identify the dideoxynucleotide incorporated, and hence the sequence of the SNP studied (the last line represents the mass obtained for the primer prior to any polymerization)

A   5'-ACGGATTACG**A**CGATCAATC-3'

5'-ACGGATTACG**G**CGATCAATC-3'

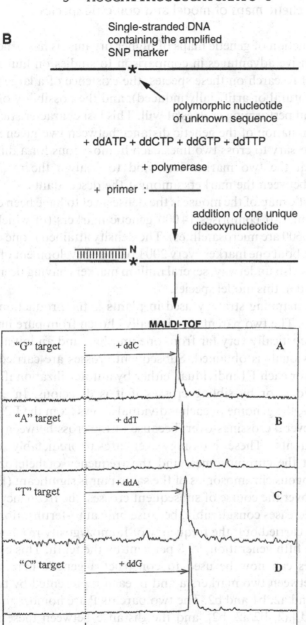

B   Single-stranded DNA
containing the amplified
SNP marker

polymorphic nucleotide
of unknown sequence

+ ddATP + ddCTP + ddGTP + ddTTP

+ polymerase

+ primer :

addition of one unique
dideoxynucleotide

MALDI-TOF

"G" target   + ddC   **A**

"A" target   + ddT   **B**

"T" target   + ddA   **C**

"C" target   + ddG   **D**

Non extended primer   **E**

5800 5900 6000 6100 6200 6300 6400 6500 6600 6700

C   Mass (m/z)

2.2.5   Genetic maps of model and domestic species

The construction of genetic maps of higher organisms has benefited from several decisive advantages in comparison to studies on humans: a long tradition of research on these species, the existence of a large number of mutants (natural or artificially induced), and the possibility of establishing lines and performing crosses at will. This last characteristic allows the easy determination of the genetic distance between two given markers: it is only necessary to cross two lines each homozygous for a different pair of alleles at the two marker loci, and to analyse the recombination frequency between the markers amongst the descendants.

The genetic map of the mouse is the densest yet to have been established for a mammal. It comprises 14 000 genetic markers (of which 3500 are genes and 6600 are microsatellites). The density attained is one marker per 0.1 cM (or about one marker every 200 kb). The development of the mouse SNP map is also underway, several million markers having been identified and mapped in this model species.

Another mapping strategy used in plants is the production of inbred generations. The two parents are initially chosen from pure lines (homozygotes), genetically very far from one another, and an initial collection of F1 individuals is obtained. Subsequent crosses are carried out independently for each F1 individual, either by autofertilization if the species is monoecious, or by sibling crosses if it is dioecious. In each of the generations, the genome of each individual shows from the F2 generation onwards several crossings-over, because of the cross between genetically distant parents. These crossings-over are unpredictably distributed throughout the chromosomes, and the frequency of heterozygosity of the homologous chromosomes of the same pair is significant (Figure 2.6). However, over the course of subsequent crosses, the frequency of heterozygosity decreases considerably, because only auto-fertilization or sibling crosses are carried out: the frequency of homozygosity reaches 93.75 per cent by the fifth generation, 99.8 per cent by the tenth. This collection of inbred lines can now be used to construct a genetic map, giving the distance between two markers a and b, each represented by two distinct alleles a1 and a2, b1 and b2. The two parents P are homozygotes [a1, b1/ a1, b1] and [a2, b2/a2, b2], and the distance between these markers is estimated by the proportion of recombinants [a1, b2/a1, b2] or [a2, b1/a2, b1] finally observed amongst the inbred lines, divided by the total number of inbred lines analysed. This approach presents several advantages: on the one hand the populations of the inbred lines can be kept

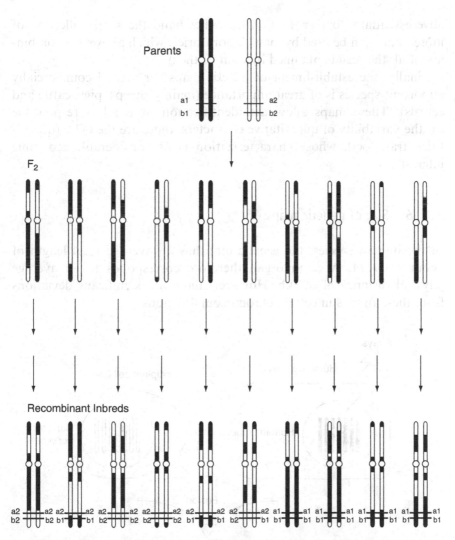

Figure 2.6 Genetic mapping by analysis of recombinant lines. The genomes of the two parents are symbolically represented by different colours, and two markers al and bl, and a2 and b2, are represented on each parental chromosome. The Fl generation inherits one of the two homologous chromosomes from each parent. The genomes of the F2 generation show an increased level of crossings-over, which declines over subsequent generations. Calculation of the distance between these two markers may be done by analysis of the recombinant lines, shown at the foot of the diagram

alive essentially forever, and on the other hand the same collection of inbred lines can be used by many laboratories, which allows the combining of all the results obtained by each of them.

Finally, the establishment of genetic maps for several commercially important species is of great importance (mainly sheeps, pigs, cattle and cereals). These maps allow the identification of the loci responsible for the variability of quantitative characters: these are the QTL (quantitive trait loci), whose characterization is of considerable economic interest.

2.2.6   Size of genetic maps

In the human species, the genetic map has an average total length of about 3500 cM. A centimorgan therefore corresponds to an average physical distance of 800 kb. However, there are significant deviations from these figures in certain chromosomal regions.

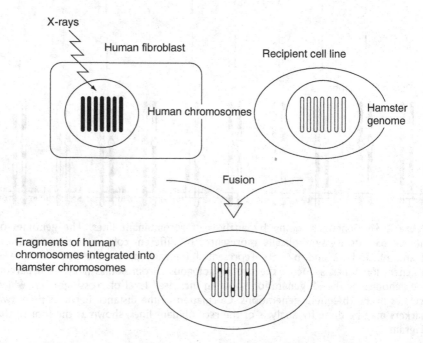

Figure 2.7 Production of radiation hybrids. Human cells are irradiated and fused with hamster cells. After fusion, fragments of human chromosomes are randomly integrated into hamster chromosomes

One remarkable feature of the human genetic map is that its size is significantly greater in females (4250 cM) than in males (2730 cM): recombination events are thus more frequent in female meioses than in male meioses (particularly in centromeric regions). This phenomenon has been observed in all mammals studied, but no satisfying explanation for it has yet been advanced.

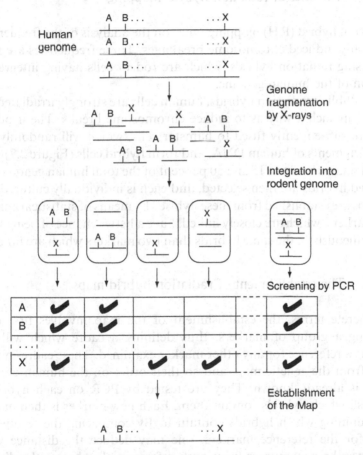

Figure 2.8 Establishing the distance between two markers using radiation hybrids. The markers A and B are initially close to each other in the genome, and distant from the marker X. After irradiation, breaks are produced less frequently between markers A and B than between these markers and X. The different fragments are integrated into hamster cells. When the markers are tested by PCR on the hybrids, one frequently finds the markers A and B together, but rarely X with A or B

In chickens by contrast, one observes no difference in genetic distance between the heterogametic sex (female, ZW) and the homogametic sex (male, ZZ), and in drosophila, crossing-over does not occur during male meiosis (total genetic distance = 0 cM).

2.3   Radiation Hybrid Maps

2.3.1   Principles of radiation hybrid mapping

Radiation hybrid (RH) mapping relies on the analysis of the frequency of artificially induced chromosome breakages. These frequencies are measured using radiation hybrids, which are rodent cells having integrated a fraction of the human genome.

To establish radiation hybrids, human cells are strongly irradiated with X-rays, in such a way as to induce chromosome breaks. The irradiated cells are subsequently fused to hamster cells, which will randomly integrate fragments of human DNA, and form hybrid cells (Figure 2.7). Each hybrid retains between 15 and 30 per cent of the total human genome. One hundred hybrids are then selected, and each is individually cultured.

Maps are established from these hybrids by means of statistical analysis: two markers, which are closely linked, have a better chance of being found simultaneously in the same hybrids than two markers which are far apart.

2.3.2   The establishment of radiation hybrid maps

In concrete terms, the establishment of the map involves first of all mapping a group of markers, thus defining a cadre which will later serve as a reference for any other marker examined. These cadre-markers come from the genetic map, and so their order on the human chromosomes is already known. They are tested by PCR on each hybrid, to establish which hybrids contain them. Each new marker is then mapped by examining which hybrids contain it. By comparing the results with those for the reference markers, one may deduce the distance which separates the new marker from each reference: the shorter the distance between two markers, the larger the number of hybrids simultaneously positive for the two markers.

The great advantage of RH mapping is that the markers do not need to be polymorphic to be mapped, because only the presence or absence of each marker is tested. This type of map is particularly useful for the mapping of EST genes (for the definition of EST, see Chapter 5), which are rarely polymorphic.

2.3.3 The first maps

Collections of radiation hybrids covering the whole genome were not developed before 1995. Two collections were used: the first was established by a collaboration between Cambridge and Généthon (GeneBridge collection), and the second was established by Stanford. These collections differ mainly in the radiation dose used. This determines the degree of fragmentation, and consequently the resolution obtained:

- High doses induce frequent chromosomal breaks, and the hybrids thus obtained are useless for mapping distant markers, because these are always separated. On the other hand, such hybrids are very useful for ordering neighbouring markers, and producing high-resolution maps. This was the option chosen when constructing the Stanford collection.

- Low doses produce hybrids useful for distant markers, but with lesser resolution. This option was chosen by the GeneBridge collection.

The GeneBridge Collection was simultaneously used by the teams of Lander and Weissenbach, the Stanford collection by that of Cox. By 1998, 30 000 ESTs had been localized using this strategy.

2.3.4 Size of maps

On RH maps, distances are given according to a specific unit: the centiRay (cR). A distance between two markers of 1 cR for an irradiation of N rad ($1\,cR^N$) corresponds to 1 per cent breaks between these two markers at this dose of radiation. It is possible to establish an average correlation between the distances obtained by this system and the physical and genetic distances. This relation varies between $1\,cR^{3000} = 200\,kb$ and $1\,cR^{9000} = 55\,kb$. These figures must nevertheless be regarded as averages, because large variations can be observed depending on the region being considered.

2.4 Conclusion

Human genetic maps so far constructed have attained a prodigious degree of resolution, even surpassing the objectives fixed in 1990 by the American programme. This density potentially allows one to undertake the localization of the genes responsible for any monogenic genetic disease, with high speed and probability of success (see Chapter 7). Nowadays several months are required to localize such a gene, whereas several years were required before the maps became available. The density of markers obtained also allows one to envisage the mapping of regions implicated in multifactorial diseases.

From a more fundamental point of view, genetic maps will allow the study of certain peculiarities of meiosis, such as the the variation of recombination frequency as a function of sex, or why some regions of the genome recombine more frequently than others. Finally, the markers ordered on the genetic map represent a very important tool for the physical mapping of the genome.

3 Physical Maps

Chromosomes are immense molecules of DNA, 20 Ångström in diameter and several centimetres long. This material is too fragile, too complex and too difficult to purify to be useable as is. The physical map of a genome has to make any desired region of that genome available in the form of a fragment of DNA, easily manipulable. The establishment of such a map begins with the construction of a DNA library representative of the genome under consideration. This stage is relatively easy to accomplish, even for large genomes like that of humans. Ordering the clones is, however, much more complex: this requires that each clone be individually identified, that its position on the chromosome from which it derives be determined, as well as its position with respect to the other clones in the library. The establishment of a precise physical map is the indispensible prelude (at least for large genomes) to work on the actual sequence: it is from this map that the minimal group of genomic clones to be sequenced will be chosen.

The establishment of a physical map relies on the data from linkage maps and partial recovery information, obtained according to the methods described in this chapter. This information allows the definition of overlap groups whose scope should in theory represent the whole genome, chromosome by chromosome.

In a physical map, distances are measured in base pairs (and multiples thereof: kilobases and megabases). These measures are thus absolute, representing physically measureable distances along chromosomes.

Genome, Transcriptome and Proteome Analysis   by Alain Bernot
© 2004 John Wiley & Sons, Ltd   ISBN 0 470 84954 1 (HB)   ISBN 0 470 84955 X (pbk)

3.1   Local Maps or Small Genomes

3.1.1   Mapping tools

The first stage in constructing a physical map is thus the construction of a genome library. The capacity of the chosen vector is a very important parameter to consider: it will determine, along with the size of the genome under study, how many clones will need to be analysed. To map a genome, it is of course preferable to use clones of as large a size as possible, so as to complete the work of mapping using the smallest possible number of clones. Bacterial vectors of moderate capacity (phages and cosmids) were generally sufficient to ensure the coverage of small genomes, this coverage eventually being completed by YACs. Conversely, the use of YACs proved indispensable for the coverage of the whole human genome with a reasonable number of clones (see the next chapter).

For small genomes, or for local maps in humans, clone order was established using overlap data for pairwise comparisons of clones. Two methods are classically used to obtain such data. The first is the establishment of restriction patterns for each clone. The comparison of these patterns allows the recognition of overlaps between clones, because if two clones overlap at least partially, a certain number of restriction fragments will have identical sizes (Figure 3.1). Another method for identifying overlaps is hybridization: two clones partially overlap if a probe derived from one hybridizes to the other.

3.1.2   Physical mapping of model species

The complete map of the genome of *E. coli* was obtained in 1987. For this, the single chromosome of the bacterium was sub-cloned into 15–20 kb fragments in a phage λ vector. The map was established in two successive stages. The first, undirected, consisted of the elucidation of the restriction patterns of 3400 clones using eight restriction enzymes. Comparison of these profiles allowed the clones to be placed into 70 groups whose size ranged from 20 to 180 kb. The second stage was to link these groups together. To do this, clones corresponding to the extremities of the previously defined groups were used to screen by hybridization the complete collection of clones in the library. By these means it was

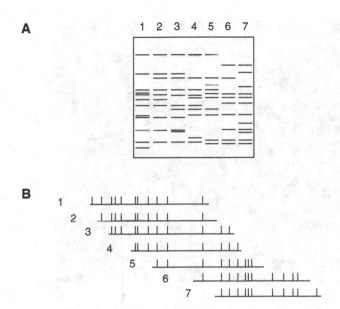

Figure 3.1 Mapping by restriction imprint. (A) The restriction map of each genomic clone (1–7) is established by enzymatic digestion and electrophoresis. (B) Comparison of the restriction maps of each clone allows the recognition of overlaps and the ordering of clones (each genomic clone is represented by an horizontal line, and the restriction sites present in each of them by vertical marks)

possible to link up the majority of the groups. The final assembly covers 99 per cent of the bacterial genome, and the whole genome is covered by a minimal selection of 381 clones. Thus it is now possible to obtain a clone covering any part of this genome, and the map was used to sequence the genome of *E. coli*.

The complete map of the genome of *S. cerevisiae* (13 Mb) was established in a similar fashion thanks to restriction pattern analysis. For larger genomes (*C. elegans*, *A. thaliana*) the establishment of maps began with the restriction analysis of cosmids, and was later completed using YACs, which made possible the detection of the junctions between the overlap groups identified by restriction mapping.

Establishing the physical map of the *Drosophila* genome was greatly facilitated by the existence of polytene chromosomes. These structures allow one to precisely map any cloned sequence by *in situ* hybridization, with a precision unequalled in other model species (Figure 3.2). The actual map relies on cosmid, P1 and YAC clones.

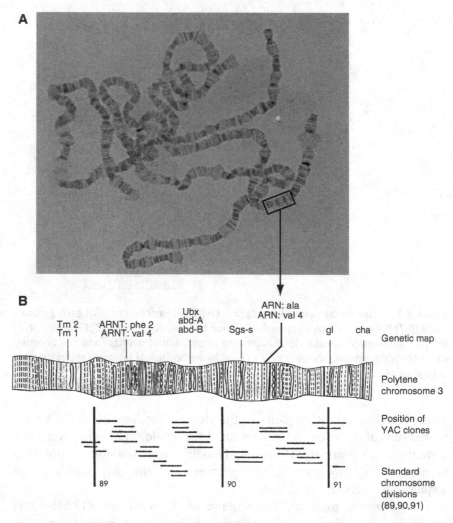

Figure 3.2 Use of polytene chromosomes for physical mapping in *Drosophila*. (A) Polytene chromosomes from drosophila salivary glands visualized by optic microscopy. The four chromosomes (X, 2, 3, 4) have been stained so as to reveal the characteristic chromosomal bands (photo from CNRS). (B) Schematic representation of a region of chromosome 3. Above are indicated the genetically mapped markers, and the horizontal bars below show the genomic clones positioned by *in situ* hybridization. The information provided by the chromosomal bands allows the establishment of the precise position of each

3.1.3 Local maps in humans

In humans, the establishment of physical maps was for a long time limited to short regions of particular interest (such as those implicated in genetic disease). Cosmids were initially the clones of choice for this type of work: later, YACs and BACs significantly improved mapping methodology.

Up to the foundation of the Human Genome Programme, clones covering a region were classically obtained by chromosome walking (Figure 3.3). This procedure involves the isolation little by little of new clones, by successive screening of a genomic library. It is evidently lengthy and painstaking, and inapplicable to the establishment of a map covering the whole of the human genome.

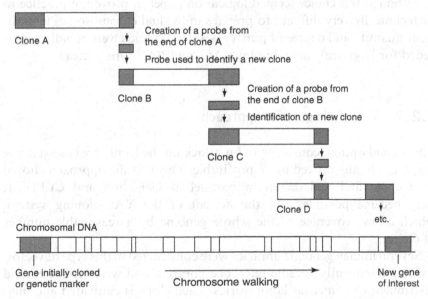

Figure 3.3 Chromosome walking. The first probe used is an already-identified segment of the genome, starting from which one wishes to explore the region which contains it. This point of entry can be a gene, already available, (for example), in cDNA form. From the first clones isolated, one can walk along the genome using the ends of the initial clones obtained as probes to re-screen the genomic DNA library

3.2   Strategies for Physical Mapping of the Human Genome

3.2.1   The chromosome-specific approach

The first attempts at large-scale mapping of the human genome were undertaken in the USA. The option chosen was the creation of chromosome-specific genomic libraries. This option was favoured because the only cloning vectors then available were cosmids, and it seemed *a priori* easier to order clones that covered only a fraction of the genome (one chromosome). Ordering a much large number of clones, covering the whole genome, seemed by contrast to be an insurmountable task. Thus American Genome Centres undertook the mapping of several human chromosomes.

Although this choice seemed logical on paper, it proved in practice to be technically very difficult to prepare individual chromosomes in sufficient quantity and degree of purity. Besides, cosmids were rapidly superceded for large-scale mapping by YAC- and BAC-type vectors.

3.2.2   The whole-genome approach

The second option, consisting of an attack on the front of whole-genome mapping, finally proved more profitable. This was the approach chosen by Cohen, and undertaken in parallel at Généthon and CEPH. It only became possible with the advent of the YAC-cloning system, which allows coverage of the whole genome by a reasonable number of clones.

Several human genomic libraries were constructed in this type of vector, which subsequently became reference libraries, and were replicated and distributed to numerous laboratories. Each clone is cultivated and analysed independently in 96-well microplates. A complete human genome library thus appears in the form of a collection of several hundred microplates, where each clone is perfectly isolated and identified.

The CEPH/Généthon library (33 000 clones) in particular, was much utilized, because of the average size of its inserts (0.9 Mb), which was superior to that of all other libraries. These clones were nicknamed 'mega-YACs'. This library was extensively analysed so as to order the clones along the genome.

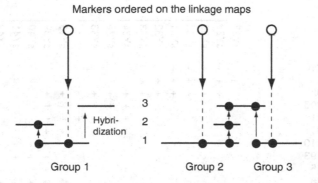

Figure 3.4 Strategies for establishing the physical map of the human genome. The top-down route (descending arrows) uses the data from linkage maps: the markers ordered on these maps (here represented by small circles) are used to identify the genomic clones which contain them, which are thereby immediately mapped. Thus, the horizontal marks on line 1 show those clones positive for the investigated markers. Identification of these clones is performed by screening the library. The bottom-up route (ascending arrows) uses local overlap data, which can be obtained by comparison of restriction profiles, by reciprocal hybridisation, or by common marker content. This may allow the identification of overlap groups (lines 1 and 3), which in some cases may be linked together (example of groups 2 and 3), thereby increasing genome coverage

Clone order was obtained by two complementary routes (Figure 3.4), known as top-down and bottom-up.

- The top-down route makes use of the data from already-established linkage maps of the genome.

- The bottom-up route uses the data from local recovery on clones, which allows the extension little-by-little of chromosome coverage, by the ordering of overlapping clones.

3.2.3   STS mapping

An STS (sequence-tagged site) is an essential tool for physical mapping. It is a short sequence uniquely represented in the genome, and easily amplifiable by PCR. An STS has no biological significance other than this, but it allows the identification within a gene library of other clones which contain it (Figure 3.5), and which therefore either cover the same region or enlarge the coverage of that region. STSs thus allow the

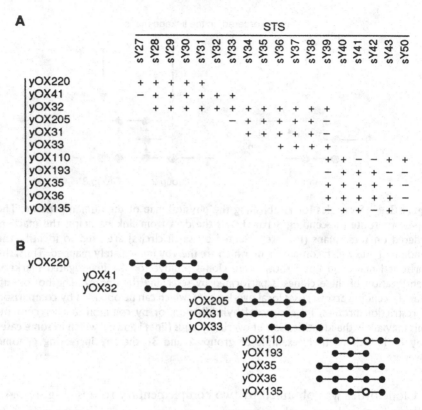

Figure 3.5 STS mapping. (A) Different YACs (yOXm) are tested by PCR with different STSs (sYn). A + sign indicates that the STS is present in the tested YAC, a − sign that it is absent. (B) Comparison of the results allows the ordering of the YACs with respect to one another. Each YAC is represented by a horizontal bar; the solid circles indicate that the YAC responds positively to the STS tested. YAC yOX110 is partially deleted: it has lost the region including the markers sY41 and sY42 (open circles)

construction of overlap groups: two genomic DNA clones are considered overlapping if they are both positive for the same STS.

The STS system presents the great advantage of being easily transposable from one laboratory to another, since it can be archived or communicated by informatic means, in the form of the sequences of the primers which define it. This allows the integration and homogenization of results from different laboratories. STSs have become in a sense the universal language of physical mapping. Some of them turn out to be associated with particularly interesting sequences: either they are in coding regions (in which case they are termed EST, for which see Chapter 5) or they are markers on the genetic map (microsatellites). In the latter case, they allow the integration of the physical and genetic maps.

The most frequent usage of an STS is the search for YACs which contain it, by library screening. This type of screening is done by PCR on the collection of clones in the library, according to the system of super-pools and pools (Figure 3.6). The results are analysed in the form of + or − responses.

3.2.4 Restriction mapping

YACs may also be characterized by their restriction pattern for different enzymes. This type of analysis was undertaken on a large scale at Généthon, in the course of characterizing the CEPH/Généthon YAC library, thanks to 20 robots constructed specially for this purpose. The DNAs of each of the clones of the YAC library were digested with three restriction enzymes. The restriction profiles were captured automatically by a camera, and an image analysis procedure allowed the detection of the restriction fragments and evaluation of their size. The information obtained was archived in an informatic form to allow the eventual detection of overlaps between different YACs.

3.2.5 Mapping by hybridization

Another method permitting the identification of overlaps between clones is hybridization of one of them to the complete library, with the aim of identifying which clones overlap at least partially with that used as probe. The other YACs in the library are themselves represented by their DNA deposited at high density onto membranes. Also placed on these membranes are DNAs from monochromosomal somatic hybrids, so as to determine from which chromosome the YAC under analysis derives.

The different stages of probe preparation and hybridization are so far as possible automated, so as to perform the maximum number of ana-lyses per day. To the same end, the results are analysed by equipment capable of automatic image capture and treatment.

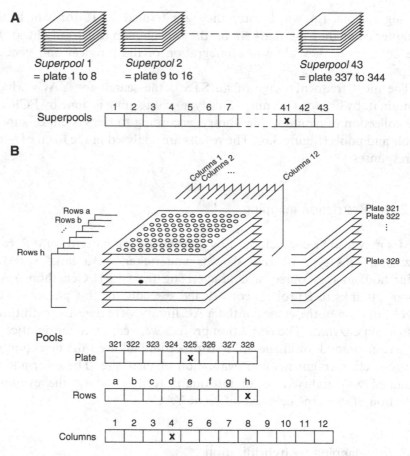

Figure 3.6 Screening a library using the system of super-pools and pools. Each super-pool corresponds to a group of clone collections from eight microplates (or 768 clones). The entire library is represented by 43 super-pools. Each super-pool is tested by PCR, so as to establish which of them responds positively to an STS of interest (in the example shown, it is super-pool no. 41). If a super-pool is positive, the identification of the positive clone within it is carried out by testing different pools from the super-pool. Each pool comprises either all the clones of each microplate, or all the clones of each column, or all the clones of each row. Positive PCR results from the pool corresponding to column 4, from the pool corresponding to row h, and from microplate 325 perfectly identify the positive clone: it will be 325_h_4. This system of screening permits a considerable reduction in the number of PCRs that have to be carried out to identify a positive clone: one reaction for each of the 43 superpools, and 28 reactions for the different pools of the superpool (eight rows, 12 columns and eight microplates). The YAC which is finally obtained is verified by one last PCR. In total, only 72 reactions are thus necessary to identify one positive clone amongst 33 000

3.3 Maps of the Human Genome

3.3.1 The CEPH/Généthon map

The first global physical map of the human genome was established at CEPH and Généthon, thanks to the mega-YAC library. This library was also used in numerous other laboratories, because of its availability to the scientific community.

Several successive publications were produced by Cohen on the mapping of the complete genome thanks to the characterization of this library. The first, appearing in 1992, made a great impact on the scientific community, because it outstripped the most optimistic prognoses on the achievement of a global map of the genome. The chosen approach particularly threw into question the American option of mapping the genome chromosome by chromosome. This was reinforced when the French team later published the first complete map of a human chromosome (chromosome 21, see below).

This map has been the subject of several successive revisions. In its latest version, top-down mapping has integrated library screening data for 2890 microsatellite markers of the genetic map constructed by Weissenbach at Généthon. Bottom-up mapping included the restriction patterns of 31 292 YACs, which allowed the establishment of linkages between 17 006 of these clones. It also included the hybridization data for 8700 clones onto the other YACs in the library, which allowed the connection of 20 890 of these YACs, and the assignment of 7209 of these to a chromosome thanks to the signals obtained from the somatic hybrids. Finally, 650 YACs were used for FISH so as to establish their genomic location.

Thanks to this enormous data collection, it was possible to establish links between a large number of YACs. These links, coupled with the location of a certain number of clones upon chromosomes, and integrated with the markers of the genetic map, allowed the achievement of a significant human genome coverage: more than 75 per cent of the human genome was covered by 225 overlap groups, with an average size of 10 Mb.

Not all these links are equally reliable, due to different technical and statistical problems. In particular, a significant amount of recombination and chimerism was detected amongst the YAC clones of the library. This means that a single YAC may contain chromosome fragments originating from two different regions, or that certain segments may have been deleted. This map must therefore be used with caution. Nevertheless it has been used successfully for the identification of a large number of genes responsible for genetic diseases.

3.3.2 The WI/MIT map

The WI/MIT team, led by Lander, also undertook a gigantic mapping project on the YACs of the CEPH/Généthon library, published at the end of 1995. This work was aided by the Genomatron, a highly automated system of PCR screening, which allowed the performance of 15 million PCRs. Thanks to this system, 10572 markers of the genetic and irradiation maps were located onto 25344 mega-YACs (top-down mapping). Additionally, the collected clones of the library were screened by 11750 STS (bottom-up mapping). The assembly of the results obtained was carried out according to strict criteria: two YACs were proposed to overlap only if they contained at least two common markers.

Integrating the collected results allowed the establishment of a first outline of a map, comprising 10850 YACs assembled in 653 overlap groups. Some of the gaps were later filled thanks to the hybridization and restriction pattern results of the CEPH/Généthon group. The final map, published by MIT and Généthon in 1995, comprised 377 overlap groups, with an average size of 8 Mb. The average distance between two markers was 200 kb, giving an extremely dense mesh size (albeit less than the objective of 100 kb set by the Genome Programme). Genome coverage was estimated at 95 per cent (Figure 3.7).

Figure 3.7 (*Opposite page*) Detail of the WI/MIT map. This page shows a fragment of the map of chromosome 14. The vertical marks represent the ordering of STSs obtained using the genetic map (to the left), using radiation hybrids (in the middle), or by YAC screening (to the right). STSs which were mapped using several tools are linked by lines. On the right of the figure the YACs are aligned as a function of the STSs they contain. This fragment is only a tiny part of the chromosome, the whole of this book would be insufficient to contain the whole map.

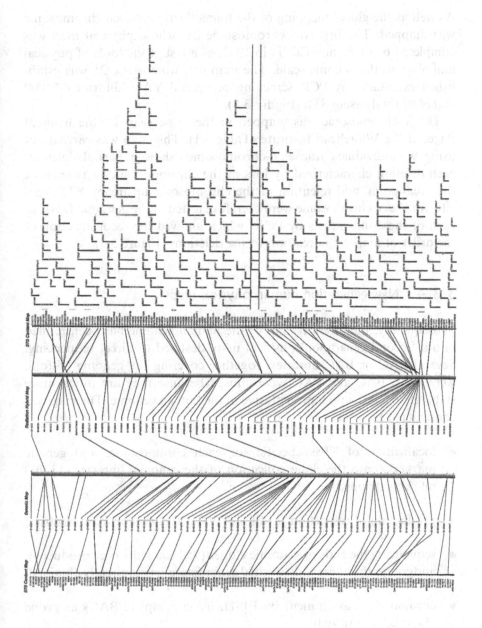

3.3.3 Chromosome-specific mapping

As well as the global mapping of the human genome, each chromosome was mapped. The first was chromosome 21, whose physical map was completed by Généthon/CEPH in 1992, as a test of the tools of physical mapping on the genome scale. The map of chromosome 21 was established essentially by PCR screening of several YAC libraries (70 000 clones in total) using STS (Figure 3.8).

The Y-chromosome was mapped in the same year, by the team of Page, at the Whitehead Institute (Table 3.1). This map was carried out using 96 individuals whose Y-chromosomes showed partial deletion, each deletion characterized by loss of the chromosome downstream of the breakpoint, and retention of the chromosome upstream. STSs specific for the Y chromosome were therefore tested in the genomic DNA of each of these individuals by PCR, which allowed the reconstruction of the order of markers in relation to the deletions (Figure 3.9).

3.3.4 New generation human physical maps

Subsequently, the physical map of the human genome has been refined into a 'second generation' physical map, destined to direct sequencing. Several PAC or BAC libraries, together covering the genome 65-fold, were used. BAC and PAC were chosen because they are much more stable than YAC, and representative of the human genome. The map was constructed by:

- localization of STSs specific for each chromosome, and genetic markers, onto the clone collection of the genomic libraries (13 700 markers mapped);

- production of restriction profiles of 300 000 clones of these libraries;

- sequencing the ends of genomic clones (750 000 sequences produced), and comparing these with finished genomic sequences already obtained;

- chromosomal assignment by FISH, using complete BACs as probe (3400 BACs utilized);

- chromosome walks, quickly performed.

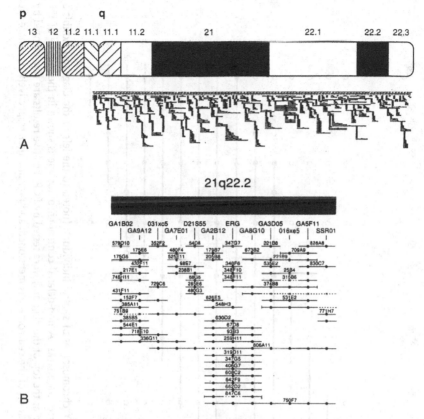

Figure 3.8 Physical map of chromosome 21. (A) Under the chromosome is shown the collection of YACs which have been mapped. (B) Detail showing how the YACs are ordered as a function of their STS content

Table 3.1 Physical mapping of chromosomes 21 and Y (1992). The depth of a map is the average number of YACs positive for a given marker; the coverage is the part of the chromosome covered by at least one YAC (nd: not done)

Chromosome	21	Y
Size (Mb)	45	50
Number of identified YACs	810	234
Average size of YAC (kb)	600	580
Number of markers mapped	198	207
Depth	10	nd
Coverage (Mb)	42	28
Average spacing of markers (kb)	220	220

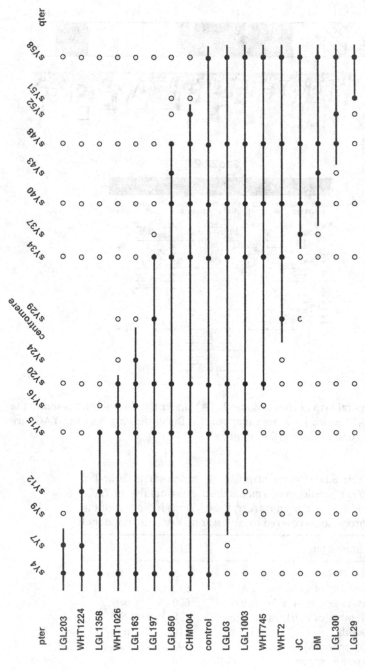

Figure 3.9 Physical map of the Y chromosome. The Y chromosomes of several individuals (noted to the left of the diagram) exhibit a partial deletion of the chromosome represented by horizontal marks. A non-deleted control chromosome is shown in the middle of the diagram. Markers specific to this chromosome (indicated at the top of the diagram) were tested by PCR: positive results are represented by solid circles, negative results by open circles. The localization of the centromere is mentioned, and pter and qter respectively designate the ends of the short and long arms of the Y chromosome

The degree of resolution obtained is very high, through a combination of the work described above and the inferences of whole-genome mapping carried out using YACs. Currently, the total map comprises about 280 groups of clones, with an average size of 10.5 Mb. It covers 96 per cent of the genome, and is very widely used within the framework of global human genome sequencing.

3.4 Conclusion

In prokaryotes, whose genomes are small, physical mapping can be rapidly performed. However, this approach is less used nowadays, and genomic sequencing of bacteria is frequently initiated without any prior physical map at all. In model species, in particular yeast, nematode or *Arabidopsis*, physical maps are nearly 100 per cent complete, principally thanks to the results obtained by hybridization, restriction profiles and utilization of STS. These maps were also the basis for complete genome sequencing.

The production of a physical map of the human genome, resting largely on the ordering of BAC and PAC, had as a goal its utilization within the framework of sequencing this genome, which is still underway (cf. the following chapter). Nevertheless it covers only 96 per cent of the genome, and its precision is inferior to the maps constructed for model organisms. This is correlated with its great complexity, some regions remaining very difficult to map physically: this is particularly true of centromeric and telomeric regions, and of regions very rich in repeated sequences. Thus, there remain several gaps on human chromosomes 20, 21 and 22 (4, 3 and 11 gaps, respectively), with the knock-on effect of the impossibility of sequencing these regions.

The mouse genome (2800 Mb) is of comparable size to that of humans (3000 Mb). Physical mapping of this genome did not begin until recently, using STS and restriction profile methods. The precision attained to date is comparable to that obtained for humans and, once again, this map is used within the framework of complete sequencing in that species.

4 Genome Sequencing

Ever since DNA was recognized to be the basis of genetic information, biologists have dreamed of sequencing the complete genome of a living organism. Effectively, to know the complete nucleotide sequence of a genome is to know all the information necessary for life (at least in theory). Such a dream remained long unattainable, for technical and financial reasons. It is only recently, with the setting up of the genome programmes, that complete genomes have been able to be entirely deciphered.

Although the complete sequence of human chromosomes is the ultimate objective of the Human Genome Programme, the size of the task to be accomplished has encouraged researchers first to sequence the genomes of model species. This allowed the evaluation of the difficulties of such operations, and the development of the tools necessary to achieve them. This stage is now over, and exhaustive sequencing of the human genome, which began in 1996, today has covered about 96 per cent of the genome.

4.1 Strategic Choices

4.1.1 Approaches used for large-scale sequencing

Two different approaches have proven efficacious for mass sequencing. The first consists of dividing the work amongst a multitude of laboratories, to which were allocated the funds necessary for sequencing the region which had been delegated to them. The entire genome sequences of the prokaryotes *B. subtilis* (46 laboratories; 1997) or *Xylella fastidiosa* (34 laboratories; 2000) were obtained in this way. The first chromosome of

Genome, Transcriptome and Proteome Analysis by Alain Bernot
© 2004 John Wiley & Sons, Ltd ISBN 0 470 84954 1 (HB) ISBN 0 470 84955 X (pbk)

an eukaryotic organism was also sequenced by this route: the sequence of chromosome III of yeast (315 kb) was obtained in 1991 by a consortium of 35 European laboratories. This success later led to the sequencing of several other yeast chromosomes.

The second approach consists of concentrating the work in Genome Centres where all the stages are carried out on a grand production scale. Such concentration permits optimal return on investment in the various stages, as well as the development of new technologies. Most of these centres concentrate on sequencing one particular genome. This option is currently dominant, and has produced significant assemblies of sequences. We may mention TIGR, which has sequenced a large number of bacterial genomes, the sequencing centre of the University of Washington at St Louis (directed by Waterston) and the Sanger Centre, who together accomplished the complete sequence of *C. elegans*, or Celera (directed by Venter), who sequenced the genome of *Drosophila* (with input from the University of California at Berkeley, Europe and Canada), and which took part in human sequencing.

The techniques used for intensive sequencing have, of course, evolved following a tendency towards more and more advanced automation. For the first chromosome of yeast, less than 10 per cent of the sequences were obtained by automatic sequencers. However, the majority of other work shows the increasing importance of automated steps, and the almost exclusive use of automatic sequencers. These machines minimize the errors introduced by human intervention, and thus increase the precision of sequencing.

4.1.2   Sequencing strategies

The strategy followed in sequencing may rest on the data from previously obtained physical and genetic maps: this is known as directed sequencing. Starting with the physical map, the first step is to choose a minimal group of clones of large fragments of genomic DNA covering the region to be sequenced (BAC, PAC, cosmid or phage), then these clones are completely sequenced (Figure 4.1).

The complete sequence of a genomic clone is obtained in the following manner: it is first sub-cloned as small fragments in an M13 or plasmid-type vector, and an initial group of sequences is obtained from a sample taken at random from these sub-clones (shotgun stage). The greater the number of clones sequenced, the more the coverage of the region extends,

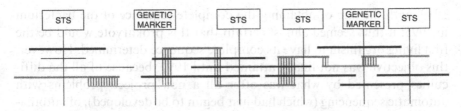

Figure 4.1 Directed sequencing. The horizontal marks show the respective localizations of different genomic DNA clones, defined by physical mapping (restriction profiles, localization of STS or genetic markers). A minimum clone collection may thus be chosen (represented in grey), sequencing of which will give the complete sequence of the region. The overlaps between successive clones are shown by vertical marks

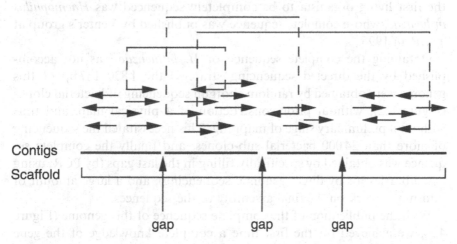

Figure 4.2 Random sequencing of a genomic clone. The arrows represent random sequences, they are assembled into contigs (thick lines) by computer programs. When two sequences are obtained from both extremities of the same genomic clone (thin and interrupted lines), these contigs can be mapped one to the other. Once ordered, these contigs represent a scaffold

but also the greater the degree of redundancy. There is a threshold level after which it becomes more economical specifically to fill in the residual gaps which remain after random sequencing. This can be done by subcloning, walking with specific oligonucleotides, PCR, or production of deletions (finishing stage).

The success of directed sequencing has not always lived up to expectations, particularly in the case of the sequencing of the *E. coli* genome. This project was 'partitioned' amongst Japanese and American laboratories in

1989, with the aim of obtaining the complete sequence of the bacterium in 1995: it thus seemed almost certain that this prokaryote would be the first living organism to have its complete sequence determined. However, this objective was not finally attained until 1997, because of all the difficulties presented by what was after all a new project: problems with automatic sequencing (which had just begun to be developed), of automation, etc. This strategy was nevertheless chosen for the sequencing of several other bacterial or eukaryotic genomes (yeast, *C. elegans*, *A. thaliana* etc.).

The significant delays experienced during the sequencing of *E. coli* threw the 'race for the complete genome' open to outsiders. In the end, the first living organism to be completely sequenced was *Haemophilus influenzae*, whose complete sequence was published by Venter's group at TIGR in 1995.

Obtaining the complete sequence of *H. influenzae* was not accomplished by the directed sequencing strategy: the 1 830 137 bp of this genome were obtained by random shotgun sequencing of bacterial clones (Figure 4.2), without prior construction of a physical map, and thus without a preliminary stage of mapping. This necessitated the sequencing of more than 24 000 bacterial sub-clones, and finally the complete sequence was obtained by specifically filling in the last gaps (by PCR, using new libraries, or by direct genomic sequencing), and a large amount of informatic work on the final assembly of the sequences.

With the publication of the complete sequence of this genome (Figure 4.3), we achieved for the first time a complete knowledge of the gene content of a living organism, as much from the point of view of its organization as of its content. This approach had previously been used with success, but on small viral genomes (for example that of phage λ, covering 48.5 kb, and obtained in 1982).

Today, random sequencing is practised in the majority of prokaryotic sequencing programmes. However it constitutes only a preliminary stage, and the remaining gaps are filled by a finishing stage, facilitated by the fact that bacterial genomes are of small size – a few megabases – and generally represented by a single chromosome. Random sequencing has subsequently been carried out on several programmes concerned with eukaryotic organims: *Drosophila*, rice and human: thus, Celera participated in sequencing the human genome by a completely random approach, while the approach followed by public laboratories relies on the genetic and physical maps achieved within the framework of the Human Genome Project.

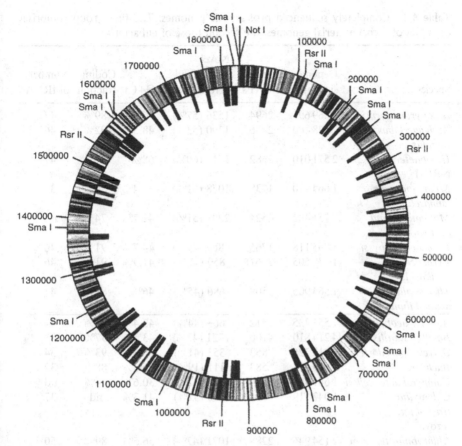

Figure 4.3 The genome of *Haemophilus influenzae*. Outer circle: restriction sites (enzymes *NotI, SmaI, RsrII*) and numbering of bases. Middle circle: ORFs, coloured according to function. Inner circle: (G+C)-rich or (A+T)-rich regions

4.1.3 Organisms sequenced

Bacteria such as *E. coli, B. subtilis, Agrobacterium tumefaciens* or *Lactococcus lactis* have been sequenced because of their importance in fundamental research, or because of their use in agriculture or in the food industry.

A large number of genomes of prokaryotic pathogens has been sequenced (Table 4.1). Priority has been given to human pathogens, because most of our illnesses are of bacterial origin. The first choices included *H. influenzae* (responsible for bronchitis, otitis), *Chlamydia trachomatis* (genital, lung and eye infections), *Chlamydia pneumoniae*

Table 4.1 Completely sequenced prokaryotic genomes. The first group comprises examples of archaebacterial genomes, the second those of eubacteria

Species	Genome size (bp)	Number of ORFs	Novel ORFs (and percentage)	(G+C)	Coding fraction	Number of tRNAs
Aeropyrum pernix	1 669 695	2694	1536 (57%)	56.3%	89%	47
Archaeoglobus fulgidus	2 178 400	2436	1290 (53%)	48.5%	92%	46
Halobacterium NRC-1	2 571 010	2682	1615 (60%)	68%	nd	47
Methanococcus jannaschii	1 664 970	1738	1078 (62%)	31.4%	88%	37
Methanosarcina acetivorans	5 751 492	4524	2298 (51%)	42.7%	74%	nd
Pyrococcus abyssi	1 765 118	1765	865 (49%)	44.7%	91%	46
Pyrococcus horikoshii	1 738 505	2061	859 (42%)	41.9%	91%	46
Thermoplasma acidophilum	1 564 905	1509	686 (45%)	46%	87%	45
Aquifex aeolicus	1 551 335	1512	663 (44%)	43.4%	94%	44
Bacillus subtilis	4 214 810	4100	1721 (42%)	43.5%	97%	88
Borrelia burgdorferi	910 725	853	353 (41%)	28.6%	93%	34
Buchnera	640 681	583	111 (19%)	26.3%	88%	32
Campylobacter jejuni	1 641 481	1654	367 (22%)	30.6%	94%	nd
Chlamydia trachomatis serovar D	1 042 519	894	290 (32%)	41.3%	nd	37
Chlorobium tepidum	2 154 946	2288	1071 (47%)	56.5%	89%	50
Deinococcus radiodurans	3 284 156	3187	1694 (53%)	66.6%	91%	49
Escherichia coli	4 639 221	4288	1632 (38%)	50.8%	88%	86
Fusobacterium nucleatum	2 174 500	2067	673 (33%)	27%	90%	47
Haemophilus influenzae	1 830 127	1743	732 (42%)	38%	87%	54
Helicobacter pylori 26695	1 667 867	1552	657 (42%)	39%	91%	36
H. pylori J99	1 643 831	1495	621 (42%)	39%	91%	36
Lactococcus lactis	2 365 589	2310	828 (36%)	35.4%	87.4%	62
Mycobacterium leprae	3 268 203	1604	nd	57.8%	49.5%	nd
Mycobacterium tuberculosis	4 411 529	3924	628 (16%)	65.6%	91%	45
Mycoplasma genitalium	580 070	470	187 (39%)	32%	88%	33

Table 4.1 (*continued*)

Species	Genome size (bp)	Number of ORFs	Novel ORFs (and percentage)	(G+C)	Coding fraction	Number of tRNAs
Mycoplasma pneumoniae	816 394	677	344 (46%)	40.0%	89%	33
Mycoplasma pulmonis	963 879	782	296 (38%)	26.6%	91.4%	29
Neisseria meningitidis (MC58)	2 272 351	2158	877 (41%)	51.5%	83%	59
Pseudomonas aeruginosa	6 264 403	5570	2549 (46%)	66.6%	89%	63
Ralstonia solanacearum	5 810 922	5129	2868 (56%)	67%	93%	58
Rickettsia prowazekii	1 111 523	834	208 (25%)	29.1%	76%	33
Streptomyces coelicolor	8 667 507	7825	nd	72.1%	89%	63
Streptococcus pneumoniae	2 160 837	2236	796 (36%)	39.7%	nd	58
Synechocystis sp.	3 573 470	3168	1766 (50%)	47.7%	87%	41
Thermoanaerobacter tengcongensis	2 689 445	2588	1094 (42%)	37.6%	87%	55
Thermotoga maritima	1 860 725	1877	863 (46%)	46%	95%	46
Treponema pallidum	1 138 006	1041	464 (47%)	52.8%	93%	44
Ureaplasma urealyticum	751 719	613	288 (47%)	25.5%	93%	nd
Vibrio cholerae	4 033 460	3885	1806 (46%)	47.5%	88%	98
Xanthomonas campestris	5 076 187	4182	1474 (32%)	65%	84%	53
Xylella fastidiosa	2 679 305	2782	1393 (50%)	52.7%	88%	49
Yersinia pestis	4 653 728	4012	nd	47.6%	83.8%	70

(pharyngitis, bronchitis, pneumonia), *V. cholerae* (cholera), *Mycoplasma genitalium* (genital tract), *Mycoplasma pneumoniae* (pneumonia), *Helicobacter pylori* (ulcers), *Mycobacterium tuberculosis* (tuberculosis), *Mycobacterium leprae* (leprosy), *Treponema pallidum* (syphilis), *Rickettsia prowazekii* (typhus), *B. burgdorferi* (Lyme disease), *Yersinia pestis* (plague), *Neisseria gonorrhoeae* (gonorrhea), *Neisseria meningitidis* (meningitis), *Listeria monocytogenes* (meningitis, miscarriages), *Campylobacter jejuni* (Guillain–Barré syndrome), *E. coli* O157:H7 (virulent *E.*

coli), *Ureaplasma urealyticum* (urogenital pathogen) and *Pseudomonas aeruginosa* (infectious in patients with cystic fibrosis, or immunodeficiency). Certain animal and plant pathogens have also been subject to such investigations, for example *X. fastidiosa*, a phytopathogen particularly of orange trees, *Xanthomonas campestris*, a pathogen of various crucifers, or *Ralstonia solanacearum*, capable of infecting more than 200 plant species.

Archaebacterial genomes have also been sequenced. Several of these species live at high temperatures, or exhibit unusual metabolisms (production of methane, reduction of sulphates, etc.). These analyses will therefore facilitate the discovery of industrially interesting enzymes.

Finally, certain bacteria have been sequenced because of their biological interests: this is the case with *Buchnera*, which lives in symbiosis with insects, and *D. radiodurans*, capable of survival in strongly irradiated environments.

Amongst the eukaryotes, the genome sequences of yeast, *S. pombe*, *A. thaliana*, the nematode and *Drosophila* are now finished (although a significant number of gaps remains in the last case). It should be noted that these organisms were chosen, because of their genome size, which is significantly smaller than that of humans (Table 1.4), as well as for their uses in the research or economic domain.

The sequencing of rice has also begun, because of its fundamental importance as a foodstuff: for 7000 years rice has been cultivated by humans, and it is currently the third most produced cereal in the world (after wheat and maize), with a global yield of 500 million tonnes in 2000. It constitutes the staple food of most of the world's population, with a consumption of the order of 100 kg per person per year in Asia (less than 10 kg in the West). Its genome size is of the order of 420 Mb, distributed over 12 pairs of chromosomes.

4.1.4   Identifying genes

The crucial problem in the analysis of genomic sequences is the identification of coding sequences. The identification of transcriptional units is greatly facilitated in prokaryotes by the possibility of identifying the promoters of genes relatively easily, by the virtual absence of introns, by the possibility of easily recognizing the open reading frames and their termination, and by the small size of the intergenic sequences (Figure 4.4).

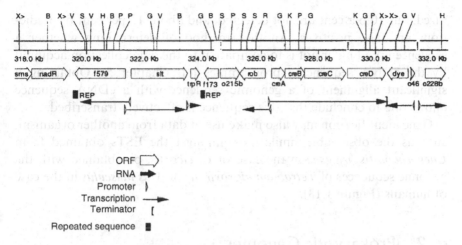

Figure 4.4 Sequence of a region of the *Escherichia coli* chromosome. The region shown is marked out in kilobases (kb). Open reading frames are represented by large arrows, oriented according to transcriptional direction. REP designates repeated sequences. Thin arrows designate the transcribed RNAs: in general, transcription initiates at a promoter (triangle) and finishes at a terminator (square bracket). When several genes are transcribed into a single messenger RNA, this constitutes an operon. The restriction sites on the map are indicated (B, *Bam*HI; V, *Eco*RV; H, *Hind*III; P, *Pvu*II; S, *Pst*I; G, *Bgl*I; K, *Kpn*I)

Two characteristics radically complicate the identification of coding sequences in eukaryotes: the division of genes into introns and exons, and the presence of intergenic regions, which are sometimes very large.

In yeast, these problems are minor, because only 5 per cent of genes are fragmented into exons, and non-coding regions are not abundant. With *C. elegans*, *D. melanogaster* or *A. thaliana*, sequence analysis is already much harder, because the majority of coding regions are fragmented, and the non-coding fraction of DNA is significant. This intron–exon fragmentation is also the rule in vertebrate genomes, where the intergenic regions are often very extensive.

The identification of transcription units remains possible thanks to informatic prediction tools capable of identifying a gene according to several criteria: the presence of an open reading frame, splicing signals and base composition. Software such as GENSCAN (for mammal sequences), GENEFINDER (nematodes), or GENIE (drosophila) integrate these parameters in a neuronal-type network or utilize Markov chains, which optimizes the probability of finding a gene. However, this type of software is not infallible: about 90 per cent of genes are recognized, but 10 per cent are not detected, and about 15 per cent of predictions are false positives. Another method of determining whether a sequence is coding or not is to compare it to the data collection acquired by the cDNA sequencing programmes (cf. Chapter 5). Obtaining a significant alignment of a genomic sequence with a cDNA sequence allows one to conclude that this sequence is effectively transcribed.

Gene identification may also make use of data from another organism, such as the observable similarities amongst the ESTs obtained from *Caenorhabditis briggsae* in the case of *C. elegans*, or obtained with the genomic sequences of *Tetraodon nigroviridis* or *Mus musculus* in the case of humans (Figure 1.13).

4.2   Prokaryotic Genomes

4.2.1   Chromosome structure

Amongst prokaryotes, the abundance of guanine and cytosine in comparison to the whole genome – termed G+C content – shows considerable variation according to species: between 26 per cent for *Buchnera* and 67 per cent for *D. radiodurans* (Table 4.1). Low values of [G+C] are often associated with a parasitic or symbiotic way of life.

Replication of the chromosome occurs in the two opposite directions diverging from the origin of replication. Each of these 'halves' is termed a replichore. With several eubacteria or archaebacteria, one observes a bias in the nucleotide frequency in each replichore. For *B. subtilis*, the ratio [G−C]/[G+C] changes sign from one side to the other of the origin of replication. For *B. burgdorferi* this change of sign is associated with an inversion of the frequency of the sequence TTGTTTTT (Figure 4.5). However, such biases are not always observed, and for certain species the location of the origin of replication remains unknown.

Genome sequencing has sometimes included that of plasmids: with *X. fastidiosa* for example, two plasmids of 1 285 and 15 158 bp were also sequenced, exhibiting respectively 2 and 64 ORF. Linear plasmids have sometimes been sequenced: *B. burgdorferi* for example contains – in addition to a linear chromosome – seven circular plasmids (from 9 to 32 kb), and 10 linear plasmids (from 17 to 56 kb). Additionally, megaplasmid sequencing has been carried out for *D. radiodurans* (Table 4.2).

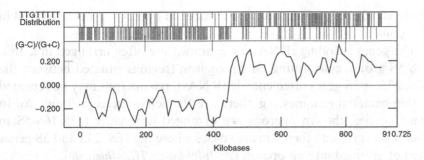

Figure 4.5   Nucleotide bias of the origin of replication. This shows the distribution of the sequences TTGTTTTT and the (G−C)/(G+C) ratio along the genome of *Borrelia burgdorferi*, around the origin of replication (central)

Table 4.2   Organization of the genome of *D. radiodurans*

	Size (in nucleotides)	Total number of genes	Genes of unknown function
Chromosome I	2 648 638	2633	1422
Chromosome II	412 348	369	183
Megaplasmid	177 466	145	65
Plasmid	45 704	40	24
Total	3 284 156	3187	1694

4.2.2   Gene organization

Globally, the coding fraction of prokaryotic genomes is elevated, of the order of 90 per cent (Table 4.1), varying between 97 (*B. subtilis*) and 49.5 per cent (*M. leprae*). The average gene size observed is of the order of 925 bp, the smallest being that of *X. fastidiosa* (799 bp), the largest that of *M. pulmonis* (1115 bp). The highest number of genes is observed in *S. coelicolor* (7825), the other extreme being presented by *M. genitalium*, which only contains 470 genes.

Amongst prokaryotes, transcription units are often organized as operons. An operon includes several genes, often implicated in the same physiological function. Each operon is transcribed from the same promoter; the genes which it contains are thus subject to indentical regulatory phenomena. A single messenger RNA is produced, and this message is sometimes subsequently cleaved into regions corresponding to each of the genes, before translation. The number of genes per operon is very variable; in *E. coli*, about a quarter of the genes are arranged in operons, whereas, in contrast, the number of operons is much reduced in *C. jejeuni*.

The genes encoding rRNA, for example, are often arranged in a 16S-23S-5S group, constituting an rrn operon (regions situated between the 16S, 23S or 5S genes often encode tRNAs). rrn operons may be repeated within bacterial genomes, e.g. there are 10 such operons in *B. subtilis*. In some species, the rrn operons are arranged differently (23S-16S-5S in *Vibrio harveyi*), and there are also cases where the 16S, 23S and 5S genes are not grouped into an operon (*A. fulgidus* or *H. influenzae*).

Generally one does not observe a preferred polarity for genes on bacterial chromosomes: genes seem to be arranged randomly on the two strands of DNA. A correlation has, however, been detected between the orientation of strongly transcribed genes and the direction of replication: for all bacteria examined to date, the direction of rRNA transcription is identical to that of replication. This might be explained by the fact that collisions between RNA and DNA polymerases could simultaneously retard both transcription and replication. In addition there are species where there is a global correlation between the transcriptional orientation of genes and the direction of their replication. This is the case, for example, with *B. subtilis*, where 75 per cent of genes are thus oriented (Figure 4.6).

Within a given species, paralogues is the name given to genes showing a high degree of similarity, and that seem to have derived from the same ancestral gene which has been duplicated during the evolution of the species. Each group of paralogues constitutes a family. Complete sequencing of bacterial genomes has allowed the identification of these families, and the frequency of paralogous genes generally increases with genome size (Table 4.3). In *B. subtilis*, for example, almost half the genome is composed of genes having at least one paralogue: this includes 284 doublets (i.e. 568 genes), 91 triplets (273 genes), 42 quadruplets (168 genes), 20 quintuplets (100 genes), etc.; and the largest family contains 77 genes. Certain families of paralogous genes have thus undergone a much larger degree of expansion than others, which suggests that genome evolution has not occurred by complete and successive duplication of an ancestral genome, but rather by independent expansions of genes individually submitted to their own evolutionary constraints.

Generally, the number of pseudogenes (mutated genes that are not transcribed or not translated) is low, of the order of 1–2 per cent: *C. jejuni*, for example, contains 20 pseudogenes. By contrast, *M. leprae*

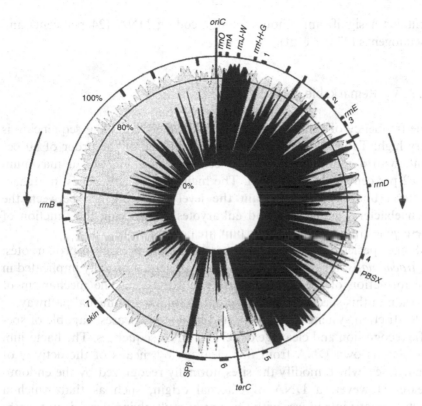

Figure 4.6 Orientation of genes along the *Bacillus subtilis* chromosome. Each replichore is represented by an arrow. Black represents the frequency of genes transcribed around the clock face. oriC, origin of replication; terC, site of termination of replication

Table 4.3 Families of paralogous genes in different prokaryotic species

Species	Number of families	Number of paralogous ORFs (and abundance in the genome)	Genome size (in megabases)
Deinococcus radiodurans	95	1 665 (52%)	3.2
Bacillus subtilis	495	1 947 (47%)	4.1
Neisseria meningitidis	234	678 (32%)	2.2
Archaeoglobus fulgidus	242	719 (30%)	2.2
Chlamydia trachomatis	58	256 (28%)	1.0
Aeropyrum pernix	185	532 (20%)	1.7
Helicobacter pylori	95	266 (16%)	1.5

contains a significant amount of non-coding DNA (24 per cent) and pseudogenes (27 per cent).

4.2.3 Remarkable genes

The frequency of totally new genes revealed by complete sequencing is very high: Table 4.1 shows this frequency to be of the order of 50 per cent, with a minimum of 19 per cent in *Buchnera*, and a maximum of 62 per cent in *M. jannaschii*. The high level of new genes in archaebacteria underlines once again the level of divergence between the archaebacteria, eubacteria and eukaryotes. Identifying the function of these genes represents an important area of research.

Some species do not contain genes habitually present in prokaryotes: *C. trachomatis*, for example, lacks the FtsZ gene, normally implicated in the formation of the septum during cell division. The mechanism of division in this species seems therefore to follow an unusual pathway.

Restriction systems are represented by endonucleases, capable of specific recognition and cleavage of certain DNA sequences. The bacterium protects its own DNA from this cleavage by means of the activity of methylases, which modify the sites normally recognized by the endonuclease. However, a DNA of external origin, such as that which a bacteriophage introduces with the aim of multiplying itself, is not methylated, and may be recognized and cleaved by the endonuclease. These restriction enzymes can thus be considered as weapons developed by bacteria in order to protect themselves from bacteriophage invasion (the cell generally does not survive this infection). Restriction enzymes generally recognize palindromic sequences, that is to say identical when they are read on each of the complementary strands. Complete genome sequencing has enabled the compilation of an inventory of restriction systems. They are found both in eubacteria and archaebacteria: 11 restriction enzymes have, for example, been identified in *H. pylori* and five in *M. jannaschii*. However in other cases, such as *C. trachomatis* or *Buchnera*, no restriction system is present.

Two-component systems allow the modulation of bacterial responses to variations in the extracellular environment. Classically these comprise a sensory protein and a regulatory protein. Fifty such systems are present in *E. coli*, 34 in *B. subtilis*, but only four in *H. influenzae*. These systems demonstrate the possible adaptations of bacteria to different environments.

DNA repair is a vital function central to the living world. It was thus particularly interesting to study it in *D. radiodurans*, which is capable of surviving in the cooling water of nuclear reactors. Numerous genes implicated in DNA repair have been identified in this species: nucleotide excision and repair, mismatch repair and several recombinational repair systems. All these genes had previously been identified in other bacteria, but *D. radiodurans* is the only species identified to date in which they have all been accumulated, often in a repetitive manner, which explains its resistance to radiation. Other resistance systems are due to the export of damaged nucleotides to the outside of the cell, and to elevated capacities of import and biosynthesis, which allow the renewal of damaged molecules (it appears that these capacities were originally acquired as an adaptation to resist dessication). By contrast, *Buchnera* contains few repair genes, and this absence of a defence mechanism is probably linked to its symbiotic lifestyle.

Studies on the metabolism of parasitic or heterotrophic bacteria show the absence of certain biosynthetic or catabolic pathways in various pathogenic species (Table 4.4). This illustrates the fact that unused genes tend to be lost. These losses are compensated for by import from the cellular host of a certain number of precursors. This may be brought about by ABC transporters (ABC = ATP binding cassette), representing one of the largest gene families yet identified in prokaryotes.

When the resources in the medium become too limited, certain bacteria activate sporulation (Figure 4.7). A septum forms, separating the cytoplasm into two unequally sized compartments – the mother cell and the prespore – each containing a chromosome derived from the last replication cycle. The prespore differentiates progressively into a spore (resistant and dormant form), a process in which the mother cell participates. When maturation is complete, the mother cell undergoes lysis, and the spore is liberated. In *B. subtilis*, chromosome analysis has allowed the demonstration that spore development requires the activation of 80 genes, regulated in a coordinate fashion.

4.2.4 Virulence

Virulence of bacterial pathogens occurs in several ways. For example, many secrete toxins, lipases or proteases (*P. aeruginosa*, *N. meningitidis*, *C. jejuni*, *V. cholerae*). The most famous example is the botulinus toxin, secreted by *Clostridium botulinum*: one gramme of this toxin could kill the entire population of Paris.

Table 4.4 Examples of metabolic pathways absent (-) in a symbiont (*Buchnera*) or various pathogens (subsequent lines)

Organism	Glycolysis	Tricarboxylic acid cycle	Amino acid synthesis	Nitrogenous base synthesis
Buchnera	+	−	+	+
Mycoplasma genitalium	+	−	−	−
Rickettsia prowazekii	−	+	−	−
Chlamydia trachomatis	+	−	+	−
Treponema pallidum	+	−	−	−
Mycobacterium leprae	−	−	+	+

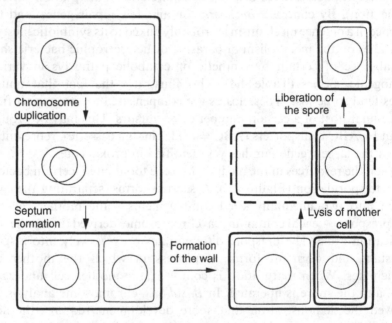

Chromosome duplication

Liberation of the spore

Septum Formation

Lysis of mother cell

Formation of the wall

Figure 4.7 Bacterial sporulation. The bacterial chromosome is duplicated, and the two chromosomes are separated by a membrane, before formation of a wall. The mother cell subsequently disappears, and the spore is liberated. Later it may germinate in a favourable environment

Resistance to the immune system is mediated by the production of immunoglobulin A proteases, observed for example in *N. gonorrhoeae*, *N. meningitidis* or *H. influenzae*. In *C. jejuni*, several genes are implicated in the synthesis of sialic acids, infrequent components of bacterial envelopes, which have been implicated in evasion of the immune system. In *H. pylori*, surface lipopolysaccharides (LPS) are very poorly immunogenic, probably because they mimic human antigens of the Lewis group.

One of the principal objectives of genome sequencing is the identification of genes implicated in pathology. In *H. pylori*, for example, the protein VacA induces the formation of acidic vacuoles in its cellular hosts, in part responsible for the appearance of ulcers. Complete sequencing has allowed the identification of three other genes whose products are similar to that of VacA, and which could also be implicated in the pathological process.

Antibiotic resistance is mediated by a considerable number of genes. Thus, the genome of *M. tuberculosis* encodes a ß-lactamase and several aminoglycoside acetyl-transferases. Note that certain species, for example *B. subtilis*, themselves encode antibiotics, which allow them to preserve their environment from invasion by other bacterial species.

Adhesion to cellular hosts is mediated by various genes encoding fimbriae, adhesins, haemagglutinins (*H. influenzae, P. aeruginosa, N. meningitidis, M. genitalium, H. pylori*, etc.). Many of these genes, which are found in surface structures, are hypervariable.

Evidence has been obtained for the presence of repeated sequences within the coding region of certain genes (Table 4.5). The number of repeats is variable, and these variations (which can induce a change in the translational reading-frame) give rise to a remarkable phenotypic diversity. This diversity allows in some cases the adaptation of these molecules to interaction with the cells of the host, in other cases evasion of the immune system. In *H. influenzae*, for example, there are tetranucleotide repeats (CAAT, GCAA and GACA) within genes coding, respectively, for an enzyme implicated in LPS synthesis, an adhesin and a glycosyl-transferase. Globally, these variations in repeat number modify transcription or translation of a surface protein involved in host–cell interaction, and hence the capacity for virulence.

4.2.5 Non-coding sequences

The non-coding domains of bacterial genomes are represented by intergenic regions, containing regulatory sequences and possibly repeated sequences, and a few rare introns. Overall, repeated sequences are much rarer in prokaryotes than in eukaryotes.

Table 4.5 Examples of intragenic repetitions implicated in pathologies

Species	Gene	Repeat	Function
Mycoplasma genitalium	*MgPa*	TAG	Adhesion to host cells
Yersinia pestis	*yadA*	GCAA	Adhesion to host cells
Neisseria	*opacity*	CTCTT	Invasion of human epithelium
Mycoplasma pneumoniae	*P1*	AACCCC	Adhesion to host cells
Staphylococcus aureus	*clf*	18 bp	Adhesion to fibrinogen
	fnb	93 pb	Adhesion to fibronectin
	cna	561 pb	Adhesion to collagen
Streptococcus	*emm*	69 pb	Escape from leucocyte phagocytosis

The distribution and abundance of these elements varies greatly between species, or even different varieties, and no general pattern has yet been discovered amongst the prokaryotes. Even a genome such as that of *M. pneumoniae*, although of very reduced size, contains 6 per cent of repeated sequences. On the other hand, the genomes of *Buchnera* or *C. jejuni* contain practically none.

In *E. coli*, intergenic sequences have an average size of 118 bp. The largest such regions attain perhaps 600 bp, and there are about 40 of them. Within about 20 of these, it has been possible to identify binding sites for proteins that regulate gene expression. The others contain either repeated sequences, or sequences whose role (if any) remains unknown.

Tandemly repeated sequences comprise a motif of one to six nucleotides, repeated from two to several dozen times. Dispersed repeated sequences do not usually contain genes. The characteristics of these regions (for *E. coli*) are summarized in Table 4.6. In some cases, their functions have been identified:

- the sites of Dam (DNA adenine methylation), 5'-GATC-3' are used by the cell to distinguish the duplicated strand from the template strand during replication or in the course of DNA repair: the original strand is distinguished from the new strand by the presence of a methyl group on the adenine;

- Chi sequences are implicated in homologous recombination;

- REP elements are palindromes which interact with HU proteins as well as with DNA gyrases, and perhaps play a role in supercoiling the bacterial chromosome.

Table 4.6 Dispersed repeated sequences in *Escherichia coli*

Type	Number per genome	Size	Fraction of the genome
IS (simple insertion sequence)	50	< 2 kb	1.5%
Rhs (recombinational hot spot)	5	6–10 kb	0.8%
REP (repetitive extragenic palindromes)	581	38 bp	0.5%
Chi (cross-over hot-spot instigator)	±1000	8 bp	0.2%
IRU (intergenic repeat unit)	19	126 bp	0.05%

In contrast, the function (if any) of other repeated sequences – such as the Rhs sequences – remains unknown.

A certain number of dispersed repeated sequences identified in *E. coli* have also been found in other species. IS elements, for example, are common, REP elements exist in *N. meningitidis*, IRU in *Salmonella typhimurium*, etc. However numerous other repeated sequences are specific to other species, and their function is unknown.

Another known peculiarity of prokaryotes is their capacity for transformation, that is to say the acquisition of a fragment of DNA sometimes followed by its integration into the chromosome in place of an homologous region. In *H. influenzae*, transformation is faciliated by a large number of sequences specific to that species: these sequences are called USS (*uptake signal sequences*), and they constitute conserved 29 bp sequences, containing a 9 bp constant region (5′-AAGTGCGGT-3′). The genome of *H. influenzae* contains 1465 USS. This high number suggests that transformation may play an important role in this bacterium. By contrast, the genome of *C. trachomatis* does not contain sequences implicated in transformation or acquisition of exogenous DNA, which is probably correlated with the fact that this parasite, isolated within the cellular host, is not in a position to acquire external DNA.

Transformation played an historically important role in the field of genetics, since it was through the work of Griffith (1928) on *Streptococcus pneumoniae* in mice and the characterization of the transforming factor by Avery (1944) that DNA was discovered to be the hereditary material.

Finally, some rare prokaryotic genes are broken up by introns. In eubacteria, examples include the gene for ribonucleotide reductase (*nrd*F), the genes encoding DNA polymerases in *B. subtilis*, and a gene encoding cytosine methyltransferase in *X. fastidiosa*. Introns are also found in archaebacteria, particularly in genes encoding tRNA and rRNA.

It is remarkable to observe introns, given the prokaryotes' overall adaptation for rapid replication. Introns have been described by certain authors as selfish DNA, duplicating itself 'for its own interests', and simply 'tolerated' to a certain degree by organims.

4.2.6 Comparative genomics

The first type of comparison is between genomes of varieties of the same species. For example, comparing the J99 and 26695 variants of *H. pylori* shows that the genome of the first contains 24 036 bp fewer than the second. The genome of J99 comprises 1495 ORFs, 89 of which are absent in 26695, and reciprocally, 117 ORFs of 26695 are missing from J66. Genome comparisons give evidence of 10 transposition or inversion events, affecting regions whose size ranges from 1 to 83 kb (Figure 4.8).

A second type of comparison can be made between the genomes of closely related species, for example those of *C. trachomatis* (a murine pathogen, 924 genes) and *C. pneumoniae* (a human pathogen, 1052 genes). Ninety-two per cent of the genes of these two species are homologous; therefore only a small number of genes specific to one or the other species. The relationship between these two genomes can also be evaluated by the identification of syntenic regions, that is to say groups of genes whose order and orientation is identical in both species. These groups seem to represent the organization which existed in the common ancestor of *C. trachomatis* and *C. pneumoniae*. There are 39 syntenic regions, and large chromosomal inversions distinguish the two species.

The organization of operons can be conserved between species, for example, about 100 operons have the same structure in *E. coli* and *B. subtilis*. This suggests that the operons concerned have been preserved ever since the common ancestor of these species, and that this organization existed before their divergence. However, if one extends this analysis to a larger number of species, such as *E. coli*, *M. genitalium*, *H. influenzae*, *Synechocystis*, *M. pneumoniae* and *H. pylori*, one finds only 14 conserved operons. Only six of these are also present in *M. jannaschii* (an archaebacterium). This suggests on the contrary that the operons do not represent an ancestral organisation: so-called conserved operons would only be cases of convergent evolution.

Another peculiarity of bacterial genomes is horizontal transfer, which is defined as DNA transmission from one species of bacterium to another (Table 4.7). This DNA may simply be acquired by the incorporation of

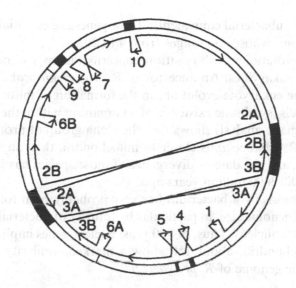

Figure 4.8 Comparison between the genomes of *Helicobacter pylori* 26695 and J99. The genome of 26695 is represented outside that of J99. The inversions and insertions in J99 are represented by extra segments

Table 4.7 Fraction of ORFs probably resulting from horizontal transfer of archaebacterial DNA into a eubacterium

Thermotoga maritima	24%
Aquifex aeolicus	16%
Bacillus subtilis	7%
Neisseria meningitidis	3%

DNA fragments present in the environment. It is thus distinct from vertical transfer, which is characterized by the transmission of genetic information from parents to offspring.

Horizontal transfer can sometimes be identified due to bias in the nucleotide frequencies which are thereby induced within the genome. Horizontal transfer has been detected in *Thermoplasma acidophilum*, with 252 genes (17 per cent of the total) probably derived from *Sulfolobus solfataricus* (an archaebacterium inhabiting the same environment). Other genes probably came from eubacteria, such as *Thiobacillus, Sulfobacillus* or *Alicyclobacillus*. Adaptation to extreme environments thus seems to have favoured genetic exchange. The genome showing the highest level of horizontal transfer is that of *T. maritima*: 451 genes – 24 per cent of ORFs – show greater similarity to archaebacterial genes than to their eubacterial equivalents. These genes are essentially grouped into 15 regions, whose size ranges from 4 to 20 kb.

Another evolutionary observation concerns the appearance of mitochondria in eukaryotes. An ancestor of *R. prowazekii* probably participated, during eukaryote evolution, in the formation of mitochondria by endosymbiosis (symbiotic existence of an organism within the cells of the host). Informatic analysis shows that the alpha group of proteobacteria, containing *R. prowazekii*, is closer to mitochondria than any other bacterial group, and the date of divergence of those species has been placed at about 1500–2000 million years ago.

Endosymbiosis by a bacterium has very probably been followed by a reduction in genome size, in particular by transfer of bacterial genes into the eukaryotic nucleus. Thus, of 300 yeast nuclear genes implicated in the life of mitochondria, about 150 show a marked similarity with genes present in the genome of *R. prowazekii*.

4.3   Genomes of Model Eukaryotes

4.3.1   Chromosome structure

Yeast was the first eukaryote to have all its chromosomes sequenced. It was hoped that this might reveal large-scale patterns within the sequences of whole chromosomes. Some general organizational features have in fact been detected:

- Some chromosomes appear as a succession of regions of about 150 kb alternatingly G+C rich or poor. This periodicity correlates with gene density, genes being more numerous in the G+C-rich regions. However, these characteristics have not been observed in all chromosomes.

- Although the complementary strands of chromosomes generally encode a similar number of genes, the number of genes encoded by each strand of chromosome II or the central region of chromosome IV is significantly different. No satisfactory explanation has yet been proposed for this observation.

- The sequence of the small chromosomes (I or VI) shows that their extremities are occupied by subtelomeric elements which are essentially non-coding. Their possible function may be to 'extend' these chromosomes so as to stabilize their structure and ensure their correct segregation during cellular division.

In *C. elegans* the genome is remarkably uniform as regards G+C content, of the order of 36 per cent. This content is practically invariant with the length of the chromosome, contrary to the observation for many other eukaryotes (particularly mammals, cf. Section 4.1). Gene density along chromosomes is slightly higher in central regions than in chromosome arms, and lower on chromosome 2.

The size of the *Drosophila* genome is of the order of 180 Mb (chromosome 4 contains only 4 Mb). Heterochromatin is very extensive in this species, covering about 60 Mb, and essentially comprising repetitive sequences, transposable elements and two blocks of ribosomal genes. Unique genes are very rare in the heterochromatin. The euchromatin covers 120 Mb, and contains the majority of the genes. It was sequenced by Celera using the whole genome shotgun: the final sequence covers 97 per cent of the genome, but 1630 gaps remain.

The genome of *A. thaliana* comprises five chromosomes whose G+C content is about 25 per cent. Two are acrocentric [chromosomes 4 (18 Mb) and 2 (20 Mb)], two submetacentric [chromosomes 3 (23 Mb) and 5 (26 Mb)], and one metacentric (chromosome 1, 29 Mb).

Telomeres are found at the ends of all eukaryotic chromosomes, comprising simple tandem repeats whose function is to maintain the integrity of chromosomal ends. The telomeric sequence is generally of the type $5'$-$C_{[1-8]}(A/T)_{[1-4]}$-$3'$ (the strand rich in cytidine is directed towards the centromere). The G-rich strand of the telomeric sequence can adopt unusual structures, including non-Watson–Crick paired hairpins, and quadruple helices. Not all telomeres are the same size, and size varies according to age: the telomere is therefore a dynamic structure. A notable exception is observed in *D. melanogaster*, where the telomeres comprise retroelements (Het-A and TART).

The centromere is a domain which appears as a constriction during the metaphase condensation of chromosomes. This region mediates the interaction between sister chromatids and initiates the assembly of the kinetochore, to which spindle microtubules attach, leading to the separation of chromosomes during division. The centromeres of several species have been cloned and sequenced; in yeast, for example, they comprise a repeated element of about 125 bp, similar in each chromosome.

4.3.2 Identification of genes

The first remarkable observation to have emerged from the analysis of organismal sequences is the high density of genes. This observation has led to an elevated estimate of the number of genes in these organisms (Table 4.8).

In *S. cerevisiae,* about 6200 genes have been identified (not counting tRNAs and genes smaller than 300 bases), about five times what had been expected. This is correlated with the fact that many of these genes have no directly observable phenotype. As a consequence they are not detectable through mutations which affect them, so they were not detected by conventional genetic techniques. Overlapping genes have been revealed, for example SMD1 and PRP38, located on

Table 4.8 Characteristics of the genomes of yeast, *S. pombe,* nematode, *drosophila,* *A. thaliana* and man. In the nematode, the gene frequency is 4.8 on autosomes, 6 on the X chromosome

	S. cerevisiae	S. pombe	Nematode	Drosophila	A. thaliana	H. Sapiens
Physical size (Mb)	13	14	100	180	125	3000
Average size of a cM (kb)	3	30	500	300	220	800
(G+C) content	38%	36%	36%	nd	41%	38%
Number of genes	6200	4900	19 100	13 600	25 500	±30 000
Coding fraction	68%	60%	27%	13%	29%	1.4%
Average number of exons per gene	1.04	1.9	5.5	4.2	5.2	8.8
Gene size (kb)	1.4	1.4	2.7	3	2.1	28
Average coding size (introns excluded)	1450	1426	1311	1497	1300	1340
Average size of exons (bp)	1450	732	218	150	250	145
Average size of introns (bp)	500	81	267	487	168	3365
Gene frequency (per kb)	2	2.5	4.8/6	9	4.5	±100
Number of tRNAs	273	174	584	284	589	535
Chromosomal location of NOR	12	3	1	X, Y	2, 4	13, 14, 15, 21, 22

opposite strands of chromosomal DNA, with overlapping 3' coding regions.

In *C. elegans*, the density of predicted genes based on the genome sequence has also exceeded previous estimates: 19 100 genes have been identified. In this species, genomic sequencing has revealed an astonishingly high number of genes located within introns of other genes, as well as of overlapping genes (Figure 4.9). Even more unexpected is the emerging evidence for the frequent organization of these genes in operons: more than 15 per cent of nematode genes are organized in this way, whereas it was previously thought that this only occured in prokaryotes.

A total of 13 600 genes have been identified in *D. melanogaster*. The surprising finding is that this figure is less than that obtained for the nematode: drosophila is a triblastic coelomate metazoan, traditionally considered to represent a more advanced stage of evolution than that seen in the nematode, which is a pseudocoelomate. A *Drosophila* also comprises 10 times more cells than a nematode, and shows more evolved behaviour (it flies!).

In *A. thaliana* 25 500 genes have been identified. This figure is higher than those of the nematode and drosophila. Yet this comparison must be carefully made: for one thing, alternative splicing is rare in *A. thaliana* (less than 5 per cent of genes, as opposed to 20–35 per cent in metazoans). For another thing, this genome has recently been subject to tetraploidization (global duplication of the genome, common in plants, cf. Section 4.3.8), which probably accounts for this high number of genes (another factor being the number of genes which encode tRNAs, higher than in any other completely sequenced eukaryote).

4.3.3 Functions of recognized or predicted genes

For a certain number of genes in all species, the analysis of the protein sequence predicted by the nucleic acid sequence allows the prediction of function (by comparison with already-known genes). Thanks to systematic sequencing, the number of genes potentially implicated in a given biological function is thus suddenly increased, much more rapidly than would have been the case had classical research methods been used. Nevertheless there remains a large fraction of genes which appear unrelated to genes of known function (from 40 to 60 per cent, according to species).

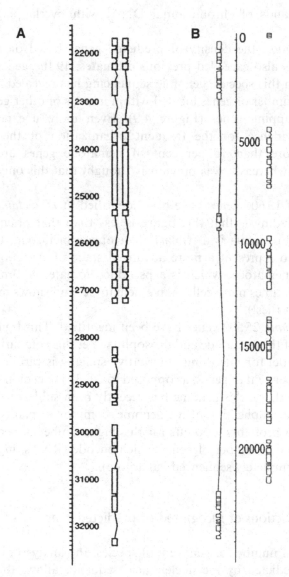

Figure 4.9 Chromosomes of the nematode. (A) Sequence of a chromosomal region from *Caenorhabditis elegans*. The region represented is marked out in base pairs. Three genes are shown as rectangles (exons) joined by broken lines (introns). The first of these shows the phenomenon of alternative splicing: the fourth exon may or may not be eliminated from the mature messenger RNA. (B) Example of genes that overlap on both strands of a chromosome. (C) Gene density in *C. elegans*, by chromosome sequenced

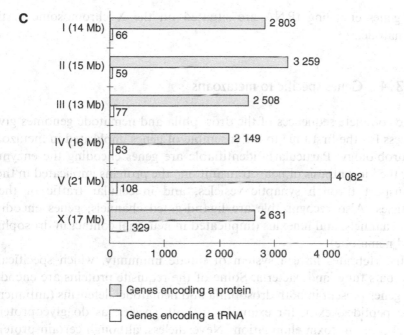

c

Chromosome	Genes encoding a protein	Genes encoding a tRNA
I (14 Mb)	2 803	66
II (15 Mb)	3 259	59
III (13 Mb)	2 508	77
IV (16 Mb)	2 149	63
V (21 Mb)	4 082	108
X (17 Mb)	2 631	329

0   1 000   2 000   3 000   4 000

☐ Genes encoding a protein

☐ Genes encoding a tRNA

Figure 4.9 *(Continued)*

Sequencing a genome can reveal unsuspected functions in an organism: in yeast, for example, a gene encoding histone H1 has been identified, whose existence had not previously been suspected. Amongst the genes found uniquely in the nematode, some encode SXC proteins, implicated in extracellular matrix interactions. In *A. thaliana* the gene hydroxynitrile lyase has been identified, which encodes an enzyme producing HCN (repellent to herbivores).

Many genes implicated in cell division are not conserved in all sequenced eukaryotes; for example, the genes encoding yeast cyclins have no equivalent in multicellular organisms. On the other hand the cyclins of drosophila, of nematodes and of vertebrates are all related.

In drosophila and the nematode, a large number of genes encode proteins implicated in the cytoskeleton (actin, tubulin, etc.) or in motility (myosin, dynein, etc.). These genes represent homologues of families present in the vertebrates.

The importance of DNA which is transcribed but not translated has been established in all species. It comprises in particular the genes encoding tRNAs, rRNAs which form the nucleolar organizers, localized in different chromosomal regions, and the 5S genes. Curiously, 40 per cent of genes encoding tRNA are situated on the X chromosome in the nematode.

4.3.4   Genes specific to metazoans

The complete sequences of the drosophila and nematode genomes gives access for the first time to the ensemble of genes implicated in metazoan neurobiology. Particularly identifiable are genes encoding the enzymes for the biosynthesis of neurotransmitters, the proteins implicated in their transport through synaptic vesicles, and in cellular traffic of these vesicles. Also recognizable are ligand-gated channels, genes encoding ion channels, and innexins (implicated in neuronal contact in drosophila and nematode).

Invertebrates have a system of innate immunity, which specifically combats fungi and bacteria. Some of the requisite proteins are encoded by genes present in both drosophila and nematode: defensins (antimicrobial peptides) exist, for example, in both species, as do glycoproteins implicated in toxin elimination. Nevertheless, although certain proteins possess immunoglobulin-type domains, there is no adaptive immune system in either drosophila or the nematode: immunogobulins and T-cell receptors have not so far been detected except in vertebrates.

The identification of genes involved in development indicates the existence of an ensemble present in all metazoa, as well as genes specific to certain lineages. In the nematode, 1600 developmental genes have been identified, and about 500 in drosophila. Homeobox proteins specifiy cellular differentiation during development by reading the DNA and regulating gene expression. Hox family genes specify cellular differentiation in space: they are organized in a gene complex, and their expression specificity along the antero-posterior axis colinearizes with their position along the chromosome. This organization is present in both drosophila and vertebrates, but in the nematode the Hox genes, whilst located on the same chromosome, are dispersed over a wide region (of the order of 3 M), separated by several hundred other genes.

Cellular adhesion is crucial in multicellular organisms; for example, several genes encoding integrins, transmembrane proteins implicated in this function, have been identified in both nematode and drosophila (there are more than 25 in vertebrates).

Several pathways of intracellular signalling are conserved between the nematode, drosophila and vertebrates (the TGF-ß and FGF pathways, tyrosine kinase receptors, Wingless/Wnt, Notch/lin-12). Some are absent in *C. elegans* (Hedghog, Toll and JAK/cytokine pathways).

Cell death is fundamental in both vertebrates and invertebrates. For example, in the nematode 131 cells are programmed to die during development. Several similar systems of apoptosis have been discovered: thus, caspases (implicated in the digestion of proteins) may be activated in both *Drosophila* and the nematode. Apoptosis may be regulated by BCL2 proteins, for which the gene exists in both species (and in vertebrates).

4.3.5 Plant genomics

The complete sequence of *A. thaliana* was obtained in 2000, covering 115 409 949 bp. Nucleolar organizers and centromeres (partially sequenced) cover about 10 Mb: genome size is therefore about 125 Mb.

The majority of genes (79 per cent) contain introns, which are often small in size. Gene density is similar on the various chromosomes, and no distribution of genes according to function is discernable.

Of the total predicted proteins, a function can be proposed for 69 per cent. Twenty-two per cent are implicated in metabolism, no doubt reflecting the importance of this function in autotrophs. The high number of genes associated with information treatment (transcriptional regulation 17 per cent, signal transduction 11 per cent) is typical of multicellular organisms.

Some genes with unexpected function, previously unknown in *A. thaliana*, have been observed; for example nodulins which are implicated in the formation of nodules (*A. thaliana* was not previously known to exhibit symbiosis with nitrogen-fixing bacteria), and the gene celA1, which encodes a cellulose synthetase similar to that of cotton.

Comparing the genes present in metazoa and *A. thaliana* reveals a certain number of similarities. Thus, there are three RNA polymerases (I, II and III), which are present in all eukaryotes. The associated proteins are similar to those known from other species in the case of polymerases II and III, but the majority of those associated with RNA polymerase I in metazoa are not identifiable (in *A. thaliana*). The number of genes implicated in transcriptional regulation is of the order of 1500, which is 1.3-times higher than in drosophila and 1.7-times higher than in the nematode, but represents a similar proportion of the total number of genes present. Nonetheless, 16 families of genes of this type are specific to

plants, and 20 families present in metazoa are absent in *A. thaliana*.
Arabidopsis also possesses DNA repair genes and recombination genes
similar to those identified in other species (several repair and recombination genes are nevertheless unique to *Arabidopsis*, whilst others are
present in metazoa but absent from the plant). The acquisition, distribution and compartmentation within the organism of substrates and
energy, vital for all organisms, is assured through 600 transport systems,
a figure comparable to that of the nematode (of the order of 700).

As in the metazoa, the cytoskeleton comprises microtubules and actin
filaments (but no intermediate filaments). These genes, like those associated with motility (kinesin, dynein, myosin), or those implicated in the
traffic and storage of vesicles, are similar to those known from metazoans or yeast. Additionally, most of the genes implicated in intracellular
activity (vesicle traffic, cell cycle) are conserved by comparison with their
metazoan homologue. There are also 150 genes for defense against
pathogens, and some of these are similar to genes identified in drosophila
and humans.

Several biological domains seem to show adaptations completely different from those of metazoa. Cellular division represents a typical
example: whilst it occurs by constriction of the cytoplasmic membrane
in yeast and metazoa, separation is effected in plants by formation of a
membrane, constructed out of vesicles originating in the Golgi appar-
atus. These two types of division are regulated by distinct proteins.
Additionally, metaphytes and metazoans have specific developmental
pathways. For example, antero-posterior differentiation is controlled
by the Hox genes in metazoa, whilst the differentiation of sepals, petals,
stamens and carpels in metaphytes is carried out by MADS box genes.
Hox and MADS genes are present in both metazoa and metaphytes,
which suggests that these shared transcription factors perform regulatory
functions that are completely distinct in the two groups.

Finally, many characteristics are obviously restricted to plants. This
may be why, in the area of transcription, a fourth RNA polymerase has
been discovered (although its role remains to be determined). This plant
has only about 40 different types of tissue, but their cellular organization
is distinctly different from that of metazoa: the presence of chloroplasts
(essential for photosynthesis), peculiar organisation of the Golgi appar-
atus, extracellular wall, importance of vacuoles, direct intercellular communication through plasmodesmata, etc.. Numerous genes encoding
cytoskeletal proteins specific to plants have been identified, and no
protein-encoding genes implicated in linking the cytoskeleton with the

transmembrane extracellular matrix have been discovered. About 420 genes are implicated in the synthesis or modification of cell walls, specific to higher plants. The redundancy of these genes is significant, apparently because of the diversity of ligands utilized. The number of proteins implicated in water transport is particularly high, a reminder of the importance of this function in plants. In the area of signalling, none of the pathways known in metazoa is represented by homologous genes in *Arabidopsis*, and their transduction follows original pathways, evoking at the same time animal and bacterial systems. Most higher plants are sessile, and environmental changes are reflected by physiological responses including growth, which may be local or transmitted by means of hormones (auxin, ethylene, abscissic acid, cytokinins etc.), which have no known equivalents in metazoa. Finally, the complete sequence has permitted the identification of all the photoreceptors, and all the proteins of photosystems I and II.

For rice PAC and BAC genomic clones have been utilized, and genetic and physical maps have been constructed in preparation for directed sequencing. This sequencing is being carried out by an international collaboration involving several laboratories all over the world. In parallel, several private companies (Monsanto and Syngeta) as well as a Chinese state laboratory have begun random sequencing of this genome. The precision of this approach is currently less than that which will be given by directed sequencing (still ongoing).

4.3.6 Homologues of genes responsible for human disease

About 20 yeast homologues of human genes whose mutant forms are responsible for genetic diseases have been identified. Despite the biology of yeast is far from directly transposable to that of humans, this organism has nevertheless become a very useful experimental model. For example, studies on a chimeric protein which includes a fragment of the CFTR gene product (this protein is implicated in the transport of Cl^-) have led to an understanding of how the $\Delta F508$ mutation (deletion of the triplet encoding phenylalanine at position 508 in the protein, causing cystic fibrosis) can be suppressed by supplementary mutations in this domain.

Additionally, 60 per cent of genes which may be implicated in human diseases are present in drosophila. Here again we encounter the problem of studies carried out on an organism very different from humans. This problem may be circumvented by producing *Drosophila* transgenic for a human gene: transgenesis of a mutated gene shows that in certain cases one may obtain in the insect a pathology similar to that seen in man. For example, this approach was taken in the case of spinocerebellar ataxia type 3 (dominant mutation in humans). Other studies have been carried out on genes responsible for cancer (27 genes conserved with man).

4.3.7   Non-coding regions

The distribution of repeated sequences in model organisms and man is shown in Table 4.9. Over the whole genome it approaches 15 per cent in *Drosophila* (because of its abundant heterochromatin), but this fraction is still very much smaller than that in man.

In *C. elegans*, there is evidence for various families of repeated sequences, both tandem and dispersed. Some of these families are preferentially located in introns and others are excluded from them. Additional location biasses are found in relation to the various autosomes and X chromosome, or between autosomal arms and central regions.

The genome of *A. thaliana* contains a small number of repeated sequences, of the order of 10 per cent (Table 4.9), and this fraction is very small by comparison to that found in other plants. In the centromeric or pericentromeric regions the numbers of transposable elements and repeated sequences is elevated, and shows a distinct distribution of different families. On the other hand the number of genes in these regions is low, and although many of these are non-functional a fraction is nevertheless transcribed, some of which correspond to genes of known function (for example ATPases, ABC transporters, enzymes of DNA replication, helicases, etc.).

Table 4.9   Frequencies of repeated sequences (in brackets: frequencies in euchromatin and heterochromatin in drosophila)

	A. thaliana	Nematode	*Drosophila*	Humans
LINE/SINE	0.5%	0.4%	4.7% (0.7 % + 13.2%)	28%
Retrovirus-type sequences	4.8%	0%	6.4% (1.5 + 16.9%)	7%
Transposon-type sequences	5.1%	5.3%	3.6% (0.7 + 9.9%)	3%
Total	10.5%	6.5%	14.9% (3.1 + 40.2%)	38%

4.3.8   Evolutionary genomics

Globally, for yeast and the nematode, about 35 per cent of genes have a human homologue. This group includes genes implicated in metabolism, synthesis and repair of DNA and RNA, protein conformation, protein trafficking and degradation. But the fraction of drosophila genes with human homologues is 50 per cent, which suggests that drosophila – despite its small number of genes – is more similar to vertebrates than either the nematode or yeast.

In yeast the degree of redundance is elevated, since a significant number of genes or regions identified on one chromosome are found to be duplicated on the same or a different chromosome. Biochemical studies show that these redundant genes encode proteins that can substitute for one another, which no doubt partly explains why alteration of a given gene does not always lead to an identifiable phenotypic modification (80 per cent of genes can be mutated without altering the survivability of this species). The fact that a genome as small as that of yeast shows such a high degree of redundance has led to the idea that the existing genome passed through a stage of tetraploidization about 100 million years ago.

A comparable observation from the total genome of *A. thaliana* is that 24 regions greater than 100 kb are duplicated (about 58 per cent of the genome). The size of these duplications suggests that *A. thaliana* too has recently passed through a tetraploid stage, followed by the loss of some of the duplicated genes. Curiously, certain regions containing telomere-type sequences are localized close to centromeric regions of *A.thaliana:* these may reflect evolutionarily recent inversions of chromosome arms.

In *C. elegans* or drosophila, a certain number of duplications have been identified, involving regions whose size ranges from several hundred bases to several tens of kilobases. This type of local duplication no doubt reflects a mechanism by means of which, during evolution, new genes have appeared through divergence from an original gene. However, these genomes do not seem to have undergone global duplication.

The complete sequences of drosophila and the nematode reveal genes which, despite being found in vertebrates or other invertebrates, are absent from one or other of these species. Genes of this type have probably been selectively eliminated in the course of evolution. By contrast, genes found only in drosophila or the nematode may represent selective inventions. The identification of genes implicated in development has for example revealed a group existing in all metazoans, but absent from the nematode and drosophila, as well as genes which are restricted to one or other of these lines.

Some genes represent the acquisition of novel functions through evolution: *Apterous*, for example, is involved in the development of limbs in drosophila and vertebrates, but not in *C elegans* – which doesn't have any. In contrast, a gene with homologues in different species may be implicated in functions that are similar but distinct in different groups. The gene *aniridia*, for example, is implicated in eye development in vertebrates; its homologue *eyeless* plays a similar role in drosophila, but in the nematode the homologue *mab-18* is implicated in the development of another sensory system, namely olfaction.

In *A. thaliana*, transfer from the mitochondrial genome to the nucleus has certainly taken place very recently, since a region of 620 kb, 99 per cent identical to the mitochondrial genome, has been identified on chromosome 2.

Finally, eukaryote–prokaryote comparisons show that the number of genes present in the former is not always greater than that in the latter; for example, the genome of *Streptomyces coelicolor* contains 7825 genes, whereas that of *S. pombe* contains only 4900.

4.4   The Human Genome

4.4.1   Human chromosomes

The total length of the human genome is about 3000 Mb, of which 2900 Mb is euchromatin. The organization of the chromosomes is shown in Table 4.10. The complete sequence of this genome (which is not finished) has involved more than 20 laboratories in six countries (USA, UK, Japan, France, Germany and China), with overall coordination by an international collaboration. The sequences were (and still are) freely and immediately accessible; they are assembled and oriented using genetic and physical mapping data, and sequence comparison. The increase in power of this international consortium is due to significant progress in technology and informatics, which have, for example, allowed overall sequencing speed to attain 1000 nucleotides per second. At the time of writing chromosomes 6, 7, 9, 10, 13, 14, 19, 20, 21, 22 and Y are complete (Table 4.11), 95.9 per cent of the genome is covered with an accuracy of 1/1000, and 91.1 per cent with an accuracy of 1/10 000.

Table 4.10 Chromosomal organization in man

Chromosome	Karyotype	Size (Mb)
1	Metacentric	266
2	Submetacentric	251
3	Metacentric	212
4	Submetacentric	197
5	Submetacentric	179
6	Submetacentric	178
7	Submetacentric	167
8	Submetacentric	145
9	Submetacentric	139
10	Submetacentric	142
11	Submetacentric	142
12	Submetacentric	141
13	Acrocentric	118
14	Acrocentric	108
15	Acrocentric	100
16	Submetacentric	94
17	Submetacentric	86
18	Submetacentric	81
19	Metacentric	68
20	Metacentric	62
21	Submetacentric	45
22	Submetacentric	48
X	Submetacentric	146
Y	Acrocentric	50

Several previous observations have been re-examined with the aid of these new data. For example, Bernardi has shown (by fractionation of genomic DNA on a caesium sulphate gradient) that the distribution of G+C in humans is far from uniform: the genome is a mosaic of regions with diverse G+C composition, termed isochores (Table 4.12). Five isochores have been recognized: the L (Light) isochores with low G+C content; L1 [(G+C) < 38 per cent] and L2 (38 per cent < (G+C) < 42 per cent), and the H (Heavy) isochores which are G+C-rich; H1 (42 per cent < (G+C) < 47 per cent), H2 (47 per cent < (G+C) < 52 per cent), and H3 [(G+C) > 52 per cent]. The H isochores correspond with the R bands obtained by Giemsa staining, and the L isochores with the G-bands. This heterogeneity is observed for all homeotherm genomes (birds and mammals), but not for poikilotherm genomes.

Another important parameter is the distribution of CpG dinucleotide islands. Over the whole genome, 50 731 islands have been identified. The

Table 4.11   Completely sequenced human chromosomes at the beginning of 2003

	14	20	21	22
Karyotype	Acrocentric	Metacentric	Sub-metacentric	Sub-metacentric
Size	87.4 Mb	59.1 Mb	33.5 Mb	33.5 Mb
Fraction of the genome	2.7%	1.8%	1.0%	1.0%
(G+C) content	40.9%	44.1%	40.9%	47.8 %
Gaps	0	4 (~320 kb)	3 (~100 kb)	11 (~150 kb)
Number of genes	1128 (of which 292 are pseudogenes)	895 (of which 168 are pseudogenes)	284 (of which 59 are pseudogenes)	679 (of which 134 are pseudogenes)

Table 4.12   [G+C] content and size of isochores

Isochore	(G+C) content	Size	Fraction of the genome	Abundance of genes
H3	(G+C) > 48%	274 kb	9%	25%
H1+H2	43% < (G+C) < 48%	203 kb	26%	27%
L	(G+C) < 43%	1079 kb	65%	48%

amount of CpG is generally linked to the number of genes present, because the islands are often located in the 5′ regions of genes. The abundance of CpG islands is thus correlated with gene density; for example, the Y-chromosome is poor in CpG (2.9 islands/Mb, 6.4 genes/Mb), while on the other hand the richest chromosome is 19 (43.4 islands/Mb, 26.8 genes/Mb).

4.4.2   Identification of genes

The human genome has 535 genes encoding tRNA, a lower number than the nematode, but higher than drosophila (Table 4.8). Nearly a quarter of these genes are located on chromosomes 1 and 6 (chromosome 22 contains none). The genes encoding 18S, 28S and 5.8S rRNA are organized in 150–200 groups of 44 kb, distributed on chromosomes 13, 14, 15, 21 and 22. Finally, the genome contains about 2000 5S ribosomal genes, a significant fraction of which are located on chromosome 1.

Protein coding genes have been predicted by (i) comparison with EST databases, (ii) comparison with complete mRNA sequences, and (iii) prediction programmes such as GENSCAN. The total number of genes in humans remains to this day uncertain, varying between 26 000 and 35 000. These figures were unexpected: they indicate that humans have only about twice as many genes as drosophila or the nematode. The average gene density is 11.1 genes/Mb, the richest chromosome being 19 (26.8 genes/Mb), and the poorest Y (2.6 genes/Mb). Some regions – of more than 500 kb – contain no genes at all; they cover 605 Mb, about 20 per cent of the genome. The longest region with no genes is a 40 Mb stretch on the Y chromosome.

Gene density varies amongst isochores: the H isochores are richer in genes than the L isochores (Table 4.12). For example, chromosomes 17, 19 and 22, which contain a lot of isochore H3, have a high gene density, and chromosomes X, 4, 18 and Y, which contain little H3, have a low gene density. Yet this distribution is not uniform: chromosome 15, which has a low amount of isochore H3, has an average gene density, while chromosome 8, which has an average amount of isochore H3, has a low gene density.

The average size of transcribed genes is 27 900 bp. The average coding fraction is 1340 bp, divided up into eight or nine exons whose average size is 145 bp (generally between 50 and 200 bp, only 42 exons are smaller than 19 bp). The average size of introns is 3365 bp. More than 35 per cent of genes exhibit alternative splicing. Overall, 12 per cent of the genome is transcribed into rRNA, mRNA, tRNA or small non-translated RNAs; only 1.4 per cent is translated.

The biggest gene is that for dystrophin, which extends over 2.4 Mb. The gene for titin also breaks records: it encodes the largest known human messenger RNA (80 780 bases) and has the largest number of exons (178), amongst which is the largest known exon (17 106 bp).

An initial evaluation of protein diversity can be obtained by the analysis of protein domains: about 51 per cent of proteins contain at least one identifiable domain, and a total of 1262 families of domains have been reported. This diversity of domain families is higher than for all other eukaryotes sequenced to date: 1035 families have been identified in *Drosophila*, 1014 in the nematode, 851 in yeast and 1010 in *Arabidopsis*.

A small number of protein domains is vertebrate-specific: only 94 of them (about 7 per cent) are not detected in any other non-vertebrate species: thus, few domains were 'invented' in vertebrates, and the majority are present in other multicellular organisms. Only one of these domains is present in a gene which encodes an enzyme, which supports the idea of an early origin for enzymatic proteins, and that the majority of such proteins are shared by all life.

Although domain invention is low amongst vertebrates, the architectural diversity of human proteins is high. An architecture is defined as the linear organization of domains along a polypeptide. Humans contain 1.8 times more architectures than drosophila or nematodes, 2.9 times more than *A. thaliana* and 5.8 times more than yeast. Another diversification of proteins is due to the expansion of certain domains: 60 per cent of domains are present in more proteins in humans than in any other eukaryote. Here we may mention FGF (30 proteins in humans, two in drosophila and the nematode), or TGF-ß (42 in humans, nine in drosophila, six in the nematode).

Some domains show a remarkable degree of expansion in humans; for example, the immunoglobulin domain (absent from *Arabidopsis* and yeast) is highly diversified. It is absent from invertebrates except for certain surface proteins, whereas in vertebrates it is found in numerous proteins of the immune system, such as immunoglobulins, T cell receptors, and proteins of the lymphocyte surface (such as histocompatibility proteins). This illustrates the diversity which a domain may achieve, concerning in this case the immune system. Similar observations have been made for domains associated with proteins of the nervous system, for example neurotrophins and their receptors, or signal proteins. Diversification in this case is linked to that of our nervous system: diversity of neuronal or glial cell types, of synaptic junctions, and of mechanisms of transmission.

Architectural diversity is also manifested by the diversity of domains with which a given domain is associated; for example, serine proteases of the trypsin type are associated in different proteins with 18 different domains in humans, but with only eight in drosophila, five in the nematode and one in yeast.

The function of about 60 per cent of proteins is either known or suspected. The principal functions of proteins are represented in Table 4.13. The most abundant are the proteins implicated in the regulation of transcription, nuclear metabolism and signal receptors. Nevertheless the function of 40 per cent remains unknown.

We have seen that the number of human genes is about twice that of drosophila or the nematode. This can be illustrated, for example, by homeotic genes, of which there are about 220 examples in humans, but only 90 in drosophila or the nematode (10 in yeast).

Table 4.13   Protein functions identified in humans

	Number	Abundance
Enzymes		
Hydrolases	1227	4.0%
Oxidoreductases	656	2.1%
Transferases	610	2.0%
Synthases - synthetases	313	1.0%
Isomerases	163	0.5%
Lyases	117	0.4%
Ligases	56	0.2%
Signal transduction		
Receptors	1543	5.0%
Regulatory proteins	988	3.2%
Kinases	868	2.8%
Signal molecules	376	1.2%
Interactions with nucleic acids		
Nucleic enzymes	2308	7.5%
Transcription factors	1850	6.0%
Various		
Proto-oncogenes	902	2.9%
Cytoskeletal proteins	876	2.8%
Cellular adhesion	577	1.9%
Transporters	533	1.7%
Proteins of the extracellular matrix	437	1.4%
Ion channels	406	1.3%
Motors	376	1.2%
Intracellular transporters	350	1.1%
Muscle proteins	296	1.0%
Immunoglobulins	264	0.9%
Transfer proteins	203	0.7%
Chaperones	159	0.5%
Viral proteins	100	0.3%
Calcium linked proteins	34	0.1%
Others	1318	4.3%

The proteomic diversity of humans is thus a function of the number of genes present, the number of protein domains and the multiplicity of their architecture. Alternative splicing also plays a role, along with diverse systems of transcriptional regulation, inter-protein interaction and post-translational modification and characteristics entering into the domain of the transcriptome or the proteome (cf. Chapters 5 and 6).

4.4.3 Repeated sequences

The human genome is the first one containing a large proportion of repeated sequences to have been sequenced: the proportion of repeated sequences is much lower in the genomes of *Arabidopsis*, the nematode or drosophila. This constitutes a significant handicap in the assembly of sequences obtained from genomic cloning because, owing to their similarity, different repeated sequences may be aligned despite coming from distinct regions.

The largest group of tandemly repeated sequences is composed of microsatellites, comprising repeats of motifs between 1 and 13 bases in length. They are frequently polymorphic, and distributed uniformly throughout the genome. These sequences have been particularly useful in establishing genetic maps (see Chapter 2). Another group of tandemly repeated sequences are the minisatellites, with motif sizes ranging between 14 and 500 bp. Such repeats may extend over 0.5–30 kb.

Dispersed repeated sequences include SINE, LINE, retroposons with LTR and DNA transposons. All derive from transposable elements, the first three via RNA, the last through DNA.

LINEs are about 6.5 kb in size (although many are truncated and their size is less than 900 bp). Generally they contain a promoter for RNA polymerase II, and two open reading frames: ORF1, which encodes an endonuclease, and ORF2, encoding reverse transcriptase. These two enzymes mediate the transposition of these elements. Three families are found in humans: LINE1, LINE2 and LINE3 (only the first is still active).

SINEs are between 100 and 300 bp in size. They contain a polymerase III promoter, but no functional gene. Their transposition uses the proteins encoded by the LINEs. About 40 families have been identifed. The

Table 4.14   Repeated sequences of the human genome

	Size	Number	Coverage of the genome (and fraction)
SINE	100–300 bp	1 500 000	360 Mb (11%)
LINE	6–8 kb	850 000	560 Mb (17%)
Retrovirus	1.5–11 kb	450 000	230 Mb (7%)
DNA transposon type sequences	80bp–3 kb	300 000	80 Mb (3%)
Total	—	—	1230 Mb (38%)

commonest SINEs are Alu sequences, MIR (mammalian wide-inter-spersed repeat), and MER (medium reiteration frequency repeats).

Retroposons with LTR (long terminal direct repeat) have a promoter and the *gag* and *pol* genes, but no *env* gene, which is necessary for infection (which suggests that they may represent an ancestral retrovirus, prior to the acquisition of *env*). These retroposons are currently inactive.

DNA transposons are flanked by inverted repeats, and encode trans-posase. Seven major classes of transposons are found in humans, but all are inactive.

LINEs and SINEs exhibit complementary distribution within the genome: Alu sequences predominate in the [G+C]-rich isochores, whilst L1 is mainly found in the [G+C]-poor isochores. LTR retroposons and DNA transposons are distributed in a uniform fashion. Overall, repeated sequences account for nearly 40 per cent of the human genome.

Some regions are very rich in dispersed repeated sequences: for example, a 525 kb section of chromosome Xp11 is 89 per cent composed of dispersed repeats. On the other hand, regions very poor in dispersed repeated sequences include that which contains the Hox genes, as well as regions very poor in genes: a 100 kb segment of chromosome 1 contains not a single gene, and only 5 per cent repeats.

4.4.4 Evolution

Amongst the proteins identified in humans to date, about 70 show no homology except with bacterial proteins: they include an epoxide hydro-lase, a glucose-6-phosphate transporter, a ß-lactamase, a hydratase and a sugar transporter. This may reflect relatively recent horizontal transfers between prokaryotes and vertebrates: about 0.1 per cent of vertebrate proteins would thus be of prokaryotic origin.

Comparing the human genome with that of the mouse allows the identification of syntenic regions between these two species (Figure 4.10), reflecting the genomic organization which existed in the common ancestor of primates and rodents, about 70 Myrs. Between these two species about 172 syntenic regions have been identified, with an average size of 15.7 Mb (the largest 84.5 Mb, the smallest 34 kb), which suggests a genomic rearrangement has occurred over these millions of years. Prac-tically all the genes of human chromosome 17 are present on mouse chromosome 11, those of human chromosome 20 are on mouse chromo-some 2, and those of human chromosome 4 are on mouse chromosome 5.

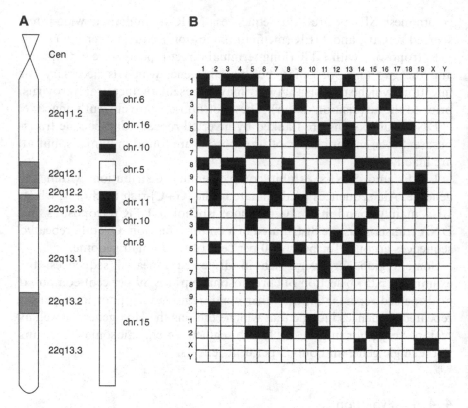

Figure 4.10   Syntenies between the genomes of man and mouse. (A) Example of human chromosome 22: regions syntenic with the mouse are shown on the right. Eight have been identified in this case, on chromosomes 5, 6, 8, 10, 11, 15 and 16 of the mouse. (B) Matrix showing the correspondence between the chromosomes of the mouse (1–19, X and Y, on the abscissa) and those of humans (1–22, X and Y, on the ordinate). Colour indicates the existence of a syntenic region between the chromosomes of these species

One important application of these syntenies is that a gene responsible for a given characteristic in the mouse has a good chance of being present in the corresponding region in humans. Thus, the gene *Pax3* was identified as responsible for the *Splotch* phenotype in the mouse, and was located on mouse chromosome 1. Separately, the region implicated in Waardenburg syndrome of humans was localized to 2q37, in a region syntenic with mouse chromosome 1. Because of the similarity between these two syndromes, the *Pax3* gene was proposed to be a candidate for this disease in humans, an hypothesis which was subsequently confirmed. Similar identification strategies have been used to search for genes implicated in multifactorial genetic diseases for which a mouse model exists.

Several authors – including Ohno – have proposed that the vertebrate genome underwent two tetraploidization stages very early in evolution. The sequence of the human genome does not support this idea: very few chromosomal regions are duplicated (in contrast to the observations made on yeast or *Arabidopsis*). Several local duplications, randomly distributed and covering 5 per cent of the genome have, however, been identified. For example, chromosomes 18 and 20 contain two regions of 36 and 28 Mb within which are 64 duplicated genes. These seem to be recent duplications, since the sequence similarity is 99 per cent.

As regards dispersed repeated sequences, the majority of these have an origin prior to the radiation of the higher mammals: sequence comparison indicates that the age of LINE1 is about 150 Myrs, that of Alu about 80 Myrs. Overall, these transposons lost their activity over the last 35–50 million years.

4.5   Other Genomes Sequenced

4.5.1   *Plasmodium falciparum*

Laveran (Nobel prize in Physiology and Medicine, 1907) identified *Plasmodium* as responsible for malaria. Four species are involved: *Plasmodium falciparum*, *P. vivax*, *P. ovale* and *P. malariae*. Between the years 1950–60, the World Health Organization (WHO) inaugurated a massive onslaught on this pathogen, focusing in particular on eliminating the mosquitos which propagate it (*Anopheles*, see below), and the use of treatments such as chloroquine or quinine. This resulted in the eradication of malaria in Europe and North Africa, but the impact of this human pathogen remains severe in other parts of the world (Table 4.15). Infected individuals manifest anemia and a deficient immune

Table 4.15   Impact of various pathogens on Man

Species	Numbers affected	Deaths per year
Plasmodium falciparum	500 million	2-3 million
Trypanosoma cruzi	16 million	50 000
Leishmania major	15 million	57 000

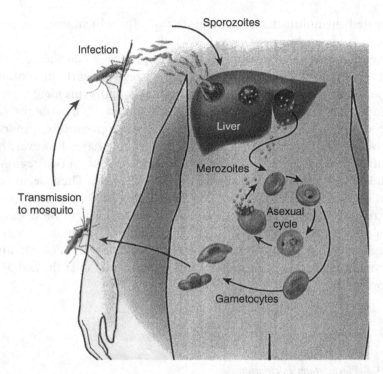

Figure 4.11 Life cycle of *P. falciparum*. Humans are infected by a mosquito which transmits the sporozoite form, and this colonizes the liver. The merozoite and tropho-zoite forms are then produced, which invade erythrocytes. Gametocyte forms may also be produced, which give rise to male and female gametes when they are absorbed by a mosquito. Fertilization gives rise to ookinetes, which colonize the mosquito and pro-duce sporozoites, the infectious form which may be reinjected into humans

system, which exposes them to other pathologies, and more than 2 million die each year, mainly in the sub-Saharan region.

The human–mosquito cycle is shown in Figure 4.11. The *Plasmodium* genome (23 Mb, 14 chromosomes) has been sequenced by an inter-national consortium. Sequencing was carried out in a random manner, but restricted to each of the chromosomes which could be separated by pulse-field gels (with the exceptions of chromosomes 6, 7 and 8, which were not separable, and were therefore sequenced together). This sequen-cing proved to be difficult and eventually took 6 years, because of the very high (A+T) content, of the order of 82%.

In total, 5279 genes have been identified (Table 4.16). In this species, the rRNA genes are not repeated in tandem, and different units are

Table 4.16 Characteristics of the genomes of P. *falciparum*, microsporidium and anopheles

	Plasmodium falciparum	Microsporidium	Anopheles
Physical size (Mb)	23	2.5	280
(G+C) content	19%	48%	35%
number of genes	5300	2000	13 700
coding fraction	52.6%	86%	7%
average number of exons per gene	2.4	1	3.5
size of genes (bp)	2 283	1 080	4 542
average size of exons (bp)	949	1 080	366
average size of introns (bp)	179	nd	1061
gene frequency (per kb)	4.3	1.0	nd
number of tRNAs	43	44	nd
Chromosomal location of NOR	chr1, 5, 7, 8, 11, 13	all chromosomes	nd

transcribed in different stages of the cycle: S units are transcribed in the mosquito, and A units in humans. The number of genes involved in metabolism is low, of the order of 14 per cent, which is probably linked to the parasitic way of life: for example, the principal source of amino acids is human haemoglobin, which is lysed into peptides and amino acids, and no amino-acid synthesis system is present in this species.

This genome contains a large number of hypervariable genes: var, vir and rif, encoding transmembrane proteins that are exposed on the cell surface. Antigenic variation of these proteins enables *Plasmodium* to escape the immune response. This variation is due to interchromosomic exchanges, transpositions between different genes, or temporally-brief activation of one gene or another. These genes are localized in telomeric regions, which are very polymorphic in *Plasmodium*.

A number of antimalarial drugs are already in use: quinine, chloroquine, sulphonamides etc. However, resistant strains of P. *falciparum* have appeared, which necessitates the development of new treatments. The availability of all the genes of this species allows one to propose a large number of new candidates. The same principle applies to vaccine development: currently only 30 *Plasmodium* proteins have been tested as vaccines, but the complete sequence suggests hundreds of new proteins, which are chosen according to their transmembrane expression, absence of polymorphism, and expression in humans.

4.5.2 *Anopheles*

Ross (Nobel Prize in Physiology and Medicine in 1902) identified the female mosquito as the transmitter of malaria. About 60 species propagate the parasite, but *Anopheles gambiae* is the most formidable: its stinger consists of one tube through which it sucks blood (*Anopheles* can thereby absorb four times its own weight from one bite), and a second through which it introduces its saliva, which includes anti-haemostatic and anti-inflammatory enzymes. It is through this latter that it may transmit the sporozoite form of *Plasmodium*: because of this propagation, *Anopheles* is currently the most dangerous animal species to humans, which has led to the complete sequencing of its genome.

Genomic libraries of *Anopheles* have been constructed starting with 330 males and 430 females. Random sequencing was carried out, the assembly of the collected sequences being complicated because of polymorphism due to the large number of different genomes employed. About 20 000 gaps currently remain in this assembly.

Over the whole genome, 13 683 genes have been identified (Table 4.16). This figure is very similar to that obtained for *Drosophila* (Table 4.8): the difference in size between their genomes is essentially due to intergenic and intronic regions which are much reduced in *Drosophila*, and the abundance of diverse families of transposons in *Anopheles*.

Analysing the collection of genes present has, for example, allowed the identification of olfactory receptors, implicated in this species' anthropophily: 276 such genes have been identified, about 2 per cent of the total. This percentage is similar to that observed in *Drosophila* and humans (humans are 3 per cent), but lower than that in the nematode (6 per cent), which lacks a visual system. In any event, it reinforces the importance of this family in the animal kingdom. Also identified were three protein families involved in insecticide resistance: Cytochromes P450, carboxylesterases and glutathione transferases, containing respectively 111, 51 and 31 genes. The immune system comprises three protein families, implicated in recognition (137 genes), modulation and transduction of signals (67 genes), and effector molecules (29 genes). Recognition is mediated through receptors specific for molecules presented by pathogens but absent in *Anopheles*. A representative example is the PGRP family (PeptidoGlycan Recognition Protein), the diversity of whose protein sequences correlates with diverse alternative splicing of their genes, thus allowing recognition of a great

PGRPLA   PGRPLC

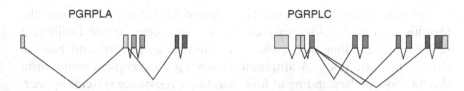

Figure 4.12 Genes encoding the pathological proteins PGRPLA and PGRPLC in *Anopheles*. Protein diversity is correlated with the possibility of alternative splicing

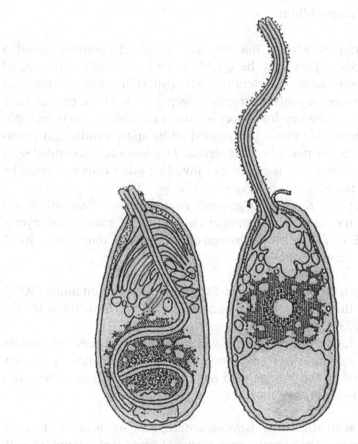

Figure 4.13 Germination of a spore of microsporidium. Extrusion of the tube allows penetration and colonization of a eukaryotic cell, of which it becomes a parasite

variety of pathogens (Figure 4.12). The effector molecules are implicated in the activation of genes encoding antifungal or antimicrobial peptides for example.

Amongst the applications which are hoped for from this sequence, the identification of the olfactory receptors in this species should facilitate a better understanding of the choice of humans as targets, and how to counter this attraction. Additionally, knowing the complete metabolism should allow understanding of how insecticide resistance systems appear, and thus how to develop new products.

4.5.3 Microsporidium

Amitochondrial eukaryotes (lacking mitochondria) comprise about a thousand species. They may be symbiotic or intracellular parasites of protozoa, invertebrates or vertebrates. Microsporidium (*Encephalitozoon cuniculi*) parasites may infect persons suffering from AIDS, or who have undergone chemotherapy for cancer or after a transplant. Their multiplication is effected by mitosis [merogony] or by spore production (sporogony); meiosis has never been observed. The spore is surrounded by a protective envelope, and is capable of invading other cells or tissues by means of an extrudable tube (Figure 4.13).

The size of the *E. cuniculi* genome is one of the smallest known amongst eukaryotes: 2.5 Mb, smaller than those of many prokaryotes (4.7 Mb for *E. coli*). The chromosomal organization is similar over its 11 chromosomes, comprising:

- a subtelomeric region covering 28 kb and always containing rRNA genes (in this species, the gene encoding 5.8S rRNA is fused to that encoding 23S).
- a central region, 90 per cent of which is coding: it does not contain repeated sequences or transposons. There are 1997 genes, very rarely interrupted by introns (only 15 have been identified), and the intergenic regions are very short, of the order of 130 bp.

Compared with other organisms sequenced, gene size is reduced: for 85 per cent of genes, the average reduction is 14 per cent compared with yeast. Dynein for example contains 4540 amino acids in *Paramecium*, 4196 in *S. pombe*, 4092 in *S. cerevisiae*, but only 3151 in microsporidium. As regards genes of known function, their analysis shows that many metabolic pathways are absent: Krebs cycle, oxidative phosphorylation, fatty acid, purine and pyrimidine synthesis, amino-acid synthesis (except

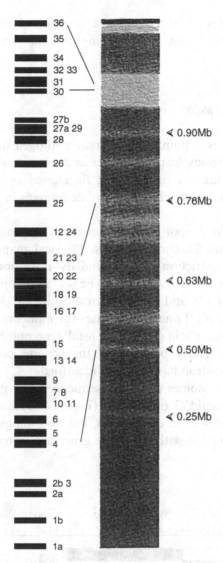

Figure 4.14 Chromosomes of *L. major* separated by pulse-field. The column on the left shows the sizes of the different chromosomes

for asparagine and serine, etc.). The function of 56 per cent of the genes remains unknown.

Analysis of several phylogenetically significant genes places this species amongst the fungi. Additionally, the presence of several genes – for example HSP70 – suggests that a mitochondrion was initially present in this species: it must have transferred several genes into the nuclear genome before disappearing. All that remains now is the mitosome.

4.5.4 *Leishmania major*

Leishmania major is a human protozoan pathogen mainly of tropical and subtropical regions (causing fevers and anemia which may lead to death). This parasite is digenetic: the flagellated extracellular form is propagated by diptera, which can inoculate humans in whom it becomes intracellular.

The WHO decided upon its sequencing. The genome extends over 35 Mb, divided into 36 chromosomes. Physical mapping was carried out by means of restriction profiles and hybridization on cosmid and PAC libraries. The chromosomes may be sorted by migration on pulse-field gels (Figure 4.14), and to date chromosomes 1, 3 and 4 have been completely sequenced. If one extrapolates the numbers of genes on these chromosomes to the whole genome, a total gene number of about 9000 may be proposed. Thirty per cent of the genome comprises repeated sequences, and no intron has yet been identified.

On all the chromosomes currently sequenced, the genes are grouped into blocks of identical 5′-3′ orientation (Figure 4.15) this is the first organization of this type seen in eukaryotes, but no biological interpretation has yet been associated with it. These genes are transcribed in a poly-

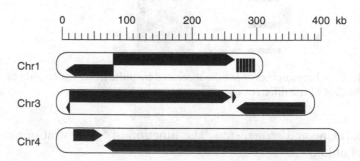

Figure 4.15 Organization of the three chromosomes of *L. major* currently sequenced. Blocks of genes with the same orientation are shown by thick bars whose ends show the transcriptional direction

cistronic manner, and a 39 nucleotide RNA is trans-spliced onto the 5' end of every messenger. Classification of genes according to the function of the proteins they encode shows that the most abundant are implicated in metabolism and protein synthesis, but for 69 per cent, the function remains unknown.

4.6 Conclusion

Over the last few years the sequencing of prokaryote genomes has proved particularly fruitful. A new bacterial genome is sequenced every 2 months, and today more than 160 complete bacterial genome sequences have been established. The great diversity of these genomes and of the genes they contain has thus been demonstrated: for about half of these genes, the function of the proteins they encode remains unknown.

Many medical spin-offs are anticipated, especially since the impact of pathogens is far from negligible: syphilis currently affects 50 million people, leprosy 15 million, a new cholera pandemic has recently appeared, and each minute tuberculosis infects 10 people. Comparing the genomes of different isolates – for example *C. trachomatis* MoPn and serovar D – may allow the identification of the genes responsible for pathogenesis. Sequence comparison between closely related species that cause very different diseases, for example *M. leprae* and *M. tuberculosis*, or *N. meningitidis* and *N. gonorrhoeae*, may also allow the identification of the genes responsible for this or that pathogenic effect.

Another approach is to use sequencing data for diagnosis or prognosis of the risk of developing infections; for example, the identification of the Ng-rep repeated sequence may be exploited in the detection of *Neisseria* contamination. Amplification of repeated sequences also allows the identification of contamination by various bacteria, including *H. influenzae*, *Bacteroides fragilis*, *H. pylori* and *N. meningitidis*.

Other projects involve eukaryote pathogens. These include *Trypanosoma cruzi*, responsible for Chagas's disease (Table 4.15).

Additional model species are being sequenced, for example *Dictyostelium discoideum* (34 Mb, six chromosomes), an organism particularly well-studied from the point of view of motility, signal transduction, differentiation and development. Complete sequencing of the mouse is also underway.

The identification of gene function in *A. thaliana* has already had spin-offs for cultivated species. For example, genes implicated in seed dispersal have been identified in *A. thaliana*, which has allowed the improvement of production in oilseed rape, as well as genes responsible for fruit maturation, which has helped the augmentation of tomato production. Additionally, two enzymes implicated in polyunsaturated lipid production have been identified in *A. thaliana*. Homologues have been identified in soya, and in a transgeneic variety of this species in which one of these genes has been suppressed, mono-unsaturated lipid abundance is increased from 25 per cent to 85 per cent, whilst that of polyunsaturates has declined from 60 per cent to 2 per cent. This example of the application of research carried out on *A. thaliana* demonstrates the potential spinoffs of the work carried out on this species.

Building on these data, the analysis of other plant genomes will be of considerable interest for humankind, because of the enormous consumption associated with them: agriculture accounts for 93 per cent of world consumption, rice, wheat and maize being the three major crops.

The size of the wheat genome is elevated because of triplication (76 per cent of its sequences are repeated). The maize genome is also large, probably because of an ancient duplication, which must have been followed by a return to the diploid state. Rice has the smallest monocotyledon genome yet discovered, and complete sequencing is underway.

Sequencing the human genome is an unprecedented programme, 25 times greater than that of drosophila, eight times greater than all the genomes sequenced. It will lead to significant medical progress, because it will allow a systematic and unbiassed approach to the genome. The identification of genes responsible for both monogenic and polyfactorial genetic diseases will thus be greatly facilitated (see Chapter 7). Overall, knowledge of this genome and the study of its genes will lead to a better understanding of their function, the stages of development, of metabolism, and of resistance to pathogens.

Finally, thanks to sequencing new concepts have been revealed, such as overlapping genes, the existence of alternative genetic codes, and messenger editing (post-transcriptional modification of mRNA). Probably this list is incomplete, and new sequences will expose new concepts.

5 Sequencing cDNA and the Transcriptome

A complete knowledge of the three-thousand million base pairs of the human genome seemed very far off at the end of the 1980s, when the genome programmes were being devised. And yet the human genome contains only a small number of sequences which encode proteins, these being *a priori* the most interesting; such sequences form the majority in a complementary DNA (cDNA) library, which represents copies of expressed messengers. Exhaustive cDNA sequencing was thus very attractive, especially because it circumvented the problem of knowing what was coding and what wasn't.

Major cDNA sequencing programmes were therefore initiated whose first, very convincing, results appeared in 1991. Subsequently a very large number of cDNA sequences has been obtained, accumulating in the data banks, and the rapid identification and localization of coding sequences was added to the list of objectives of the Genome Programe in 1993. Not all of these sequences have been archived in the public databases, because this type of programme has raised the problem of industrial and intellectual property in relation to the data produced in research on the human genome.

5.1 Strategies of cDNA Sequencing

5.1.1 Posing the problem

Eukaryotic genomes contain only a small fraction of coding sequences (see Chapter 1). The messenger RNA population of a given cell is, on the

Genome, Transcriptome and Proteome Analysis by Alain Bernot
© 2004 John Wiley & Sons, Ltd ISBN 0 470 84954 1 (HB) ISBN 0 470 84955 X (pbk)

other hand, considerably enriched in coding sequences, because it represents those destined to be translated into proteins.

However, all genes are not uniformly or permanently transcribed in all tissues: each tissue expresses only about 15 000 genes. Of these, only a portion (about 10 000) is constitutively expressed by all categories of cells: these are the so-called 'housekeeping' genes, which direct the basic functions common to all cells. The remaining several thousand messengers are specific to the tissue and/or stage of development under consideration.

Additionally, the level of expression of genes in a given population is quite variable. Some messenger RNAs are abundantly represented, at more than 1000 copies per cell (up to 10 000 copies for certain very specialised tissues). Other messengers are rare, or even very rare (up to a few dozen per cell).

A cDNA sequencing programme consists of characterizing the totality of messengers expressed by a cell, starting from a collection of independent cDNAs. It follows from the arguments above that the collection of sequences obtained will be biased in proportion to the representation of corresponding sequences in the genome.

5.1.2 Production and sequencing of cDNA

The first stage in a cDNA sequencing programme is obviously the construction of a cDNA library. This is a technique currently practised by numerous laboratories (Figure 5.1). The insert in each clone in a cDNA bank is sequenced using a primer homologous to a vector sequence situated close to the cloning site. This allows the use of the same primer for all the clones, thus facilitating large-scale sequencing.

By this method one obtains several hundred nucleotides from each extremity of the cloned cDNA. The information thus obtained is therefore partial in terms of the whole cDNA, which may be several kilobases long. Nevertheless, this information suffices to characterize each clone specifically. It is not practical to determine precisely the complete sequence of each cDNA, because this would require a finishing step difficult to implement on a large scale. Partial cDNA sequences obtained according to this strategy are termed ESTs (expressed sequenced tags).

It is appropriate to emphasize the originality of this method in relation to those employed by conventional laboratories, where one cDNA can by itself form the object of an entire research programme. Here we

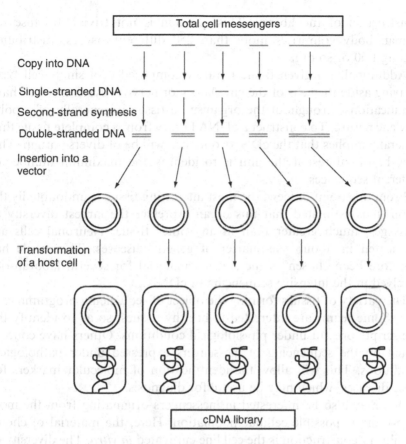

Figure 5.1 Construction of a cDNA library. Starting with a tissue or from cultured cells, poly-adenylated messenger RNAs are purified. These molecules are copied into double-stranded complementary DNA, cloned into a bacterial vector, and introduced into *Escherichia coli*. Each clone that is obtained contains a unique cDNA inserted into a vector. This DNA may be purified after a period of bacterial growth, and the cDNA is thus obtained in sufficient quantity to be sequenced

are concerned with a completely different approach: every clone is succinctly characterized, so as to allow the analysis of the largest possible number.

5.1.3 Choice of tissue of origin

The first question which occurs at the outset of a cDNA sequencing programme is the choice of the biological material to be used for the construction of the library. This question is not trivial, because the human body comprises more than 250 different tissues, distributed amongst 50 or so organs.

Additionally, a given tissue is rarely composed of a single cell type. Leaving aside the cells of the circulatory or nervous system, which have ramifications throughout the organism, a tissue is generally of a poly-phyletic nature. To construct a cDNA library from a complete tissue thus generally implies that the cDNAs obtained will be of diverse origin. This may be of interest if the aim is to identify the maximum number of different sequences.

From this point of view, the most interesting tissue is undoubtedly the brain. It is estimated that this organ expresses the largest diversity of messages, much greater than in any other tissue. Neuronal cells are implicated in about one-quarter of genetic diseases. This tissue has therefore been chosen as the starting material for several laboratories involved in the intensive sequencing of cDNA.

The variety of tissues forming the object of sequencing programmes is vast. Some teams are interested in healthy tissues, so as to identify the transcripts present under physiological conditions. Others have concen-trated on the sequencing of messengers expressed under pathological conditions. This will allow the identification of molecular markers for these diseases, which may be useful for diagnosis.

One may also be interested in messengers originating from the most homogenous possible cellular population. Here, the material of choice for library construction is the cell line cultivated *in vitro*. The diversity of messages is thus lower, but it represents precisely the collection of genes expressed in the chosen cell type.

5.1.4   Large-scale sequencing of cDNA

With the exception of clone-library construction, which relies on purely manual laboratory techniques, the simplicity of this strategy lends itself quite easily to automation, which has been developed by the majority of the laboratories involved in these programmes. Bacterial clones may be ordered on microplates by robot; PCR and sequencing reactions are also more easily automatable, because the primers necessary for both these reactions are the same for all clones from the same bank, and the sequences are obtained by automatic sequencers.

EST analysis is carried out by means of informatic programmes executed with the minimum of human intervention. Identical or overlapping sequences are grouped together. The collection of sequences is then compared with the known genes deposited in databanks, so as to determine which correspond to a known gene, in man or other species. Finally, the ESTs are deposited in a public databank.

The accumulation of ESTs has inspired the creation of an international databank, called dbEST, specially devoted to this type of sequences. The creation of such a resource was justified in part because of the fragmentary character of this type of sequence, and in part because of the rapidity with which they were produced. On its creation, in summer 1993, dbEST contained 22 500 ESTs, of which 14 500 were human; in July 2004, it contained 22 570 000, of which 5 650 000 were human.

5.2 The Economic Stakes

5.2.1 Data protection

The first results of intensive cDNA sequencing programmes showed that it was possible to progress rapidly towards information (albeit partial) concerning a large number of human genes. Some of those responsible for these programmes thus tried to patent these sequences, so as to reserve the benefits of eventual utilizations which the study of these genes might eventually produce.

Thus the NIH, which was developing such a programme, tried to patent its ESTs. This patent application was rejected, because the fragmentary nature of the information in the partial sequences (with no indication about the biological role of the corresponding protein) was not considered to merit the title of an invention.

This application nevertheless created a great stir in the international scientific community, which judged it unacceptable that patents be applied for at this stage of genome exploration, which could lead to blockage of whole sections of research.

5.2.2 The involvement of large pharmaceutical companies

The possibility of sequencing a large number of human genes relatively easily, and the important therapeutic – and thus financial – opportunities thus revealed, spurred several pharmaceutical companies to join the cDNA 'race'. Such operations were confined to the USA, because of the availability of significant capital for investment in high-risk enterprises. Considerable sums were invested, which led to the development of real 'sequencing factories'. Thus TIGR/HGS, Incyte Pharmaceuticals and Millenium have concentrated their very great resources to the extensive sequencing of cDNA.

The impossibility of protecting partial sequences by patent applications had the consequence that the data produced by these private companies were jealously kept secret, so as to protect any application which might be derived from them. Also these companies have accumulated hundreds of thousands of sequences which remain confidential and inaccessible to the scientific community. This attitude is not to be totally unexpected from private groups which invest considerable sums in research programmes, from which they obviously wish to profit. Nevertheless it is counter to the habits of the scientific community, within which publication of research for the benefit of all is the rule.

5.2.3   The 'coining' of partial sequences

These companies hope to profit from their investments by negotiating for access to their data at high prices. TIGR/HGS allows access to its partial sequence databases on condition of placing an exclusivity contract on any commercial application derived from the use of its sequences. Another strategy is not to control database access case-by-case, but rather to sell it 'en bloc'. In this case, the rights to patents or spin-offs that may be obtained from the sequences are lost, but the benefits follow immediately. TIGR/HGS, Millenium and Incyte have sold their databases to several large pharmaceutical companies in this way.

However, the blockade imposed on these data eventually had a happy consequence: the involvement of another private pharmaceutical company in the cDNA race. Merck began a vast cDNA sequencing project in 1994 whose results are accessible to all, with no conditions. This decision was taken in part to counter the TIGR/Incyte monopoly, and in part so that Merck would be well-placed to exploit these data. Additionally, this operation gave Merck excellent publicity.

To carry out this project, Merck financed public laboratories already involved in intensive sequencing, especially the University of Washington at St Louis. The first 15 000 ESTs were made public in February 1995, showing the effectiveness of the methodology. Subsequently the sequences produced were placed in public databases about 48 h after validation, at a rate of around 4000–6000 ESTs per week.

5.2.4   The cDNA race

Merck, Millenium, Incyte and TIGR/HGS have thus assumed the leadership in the area of intensive cDNA sequencing. With these companies, this type of programme has reached an even higher scale: the cDNA libraries constructed number hundreds, the numbers of sequences analysed are in the tens of thousands, and the total information collected runs to megabases.

Incyte and TIGR/HGS have prepared several hundred cDNA libraries from different tissues. The Incyte database probably contains 3 million ESTs, and more than 850 000 ESTs have been produced by TIGR/HGS. Merck has recovered from its original delay, and has published 480 000 human ESTs. Each of these programmes surpasses the total collection of results by all the other laboratories implicated in cDNA sequencing. Merck has subsequently abandoned this project, but other programmes have persevered along this route, with the setting up in the USA of the IMAGE consortium (Integrated Analysis of Genomes and their Expression), and the CGAP (Cancer Genome Anatomy Project).

5.3   The Analysis of cDNA Sequences

5.3.1   Assembly of partial sequences

The considerable number of ESTs produced by the various programmes offers the possibility of reconstituting complete sequences starting from several partial sequences. This may be done case by case, by any scientist interested in a particular gene for which only partial information is available (Figure 5.2).

This approach has also been practised by TIGR/HGS, which in 1995 compared its 174 500 ESTs with 118 400 present in the database. Where an overlap of several sequences was detected, these were assembled into a consensus sequence termed a THC (tentative human consensus sequence). TIGR/HGS thus created 29 600 THCs, which in principle cor-

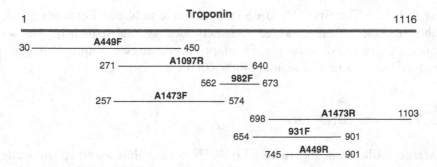

Figure 5.2 Assembly of partial cDNA sequences. Seven partial sequences encoding Troponin T here allow the reconstruction of the entire mature transcript sequence

respond to that number of genes. No assembly was possible for 58 400 other ESTs, which remain as singletons. Not all singletons will represent different genes, because two partial sequences may represent distant and non-overlapping regions of the same gene.

The diversity of sequences obtained is therefore very large, and immediately justifies the random sequencing of clones, because most of the information obtained will be original. A similarly low degree of redundance has been observed by all laboratories involved in this type of programme.

5.3.2   Identification of new genes

The major observation which emerged from the analysis of cDNA sequences by these programmes was the considerable number of new genes identified. The 80 000 entities identifed by TIGR/HGS (THC and singletons) were compared with 4550 complete gene sequences already present in the databases. A mere 10 200 corresponded to any known gene. All other laboratories observed that the fraction of unknown sequences always exceeded 50 per cent.

Some sequences may be similar to an already-known sequence, without there being absolute identity between them. Such a cDNA represents a new member of a gene family, and in this way it has been possible to enlarge the size of numerous gene families. It has also been possible to identify gene families of which only a single member was previously known. Even amongst the sequences which do not correspond to any known gene, it is possible to establish groupings defining new families: several hundred new families have thus been identified.

Amongst the new sequences a certain number are similar to genes already known from other species, but whose existence in man was unsuspected. Analysis of sequences obtained from a given tissue has also revealed many unexpected examples of specificity of expression. Just as with genomic sequencing, this type of programme therefore underlines our ignorance of many biological phenomena, and considerably enlarges our insight into cellular physiology.

5.3.3 The transcriptional map of the human genome

One of the most important outcomes which followed the increased power of cDNA sequencing was the integration of a transcriptional map with the human physical and genetic genomic maps. On the one hand this allowed – in conjunction with human genome sequencing – the identification of coding sequences. On the other, this map has been a powerful tool for the identification of genes responsible for genetic diseases based on their genomic location (cf. Chapter 7).

Locating an EST within the genome relies on PCR: starting with the cDNA partial sequence one chooses a pair of primers, which are used to identify genomic clones which contain them, and to map the corresponding gene.

The original technique used to map an EST was chromosomal assignment using mono-chromosomic somatic cell hybrids. However only mediocre resolution was obtained, because it could only be placed at the chromosomal scale. Better mapping is obtained using radiation hybrids, which allows the mapping of non-polymorphic sequences such as ESTs (cf. Section 2.3). In this method better resolution is obtained (of the order of the megabase). Several large centres have used this mapping tool: WI/MIT, the Stanford University, the Sanger Centre and Généthon (cf. Section 2.3). By 1998 this had achieved the localization of about 30 000 human genes.

EST sequences are nowadays also used to identify coding regions within the human genome, of which the complete sequencing is still ongoing.

5.3.4 Identification of genes responsible for genetic diseases

Even without mapping, cDNA sequencing allows the identification of genes responsible for genetic diseases. Thus, following the identification of the gene *hMSH2* as responsible for a form of colon cancer, research into sequences similar to this gene allowed the identification amongst ESTs of a second gene, *hMLH1*, also implicated in the appearance of this cancer. In the same way, a partial sequence of a gene involved in predisposition to Alzheimer's disease (*Presenilin* I) allowed the identification of a second gene for predisposition to this disease (*Presenilin* II).

Finally, other potentially interesting genes are those specifically expressed in cancer cells but not in the corresponding healthy tissue. Such genes may be used as markers of these pathological states, and represent candidates for being the origin of these diseases.

5.3.5   cDNA programmes in animal and plant species

Several cDNA programmes are underway for other species, both animal and plant. Table 5.1 summarizes the largest numbers of partial sequences deposited in dbEST, arranged by species. In all cases, the number of new sequences identified is large. As with man, these sequences are used to identify coding regions (exons and genes) for those species where the genome has been completely sequenced.

Table 5.1 Number of partial sequences deposited in the dbEST database. Only the 10 most abundant species are shown (August 2003)

Species	Number of EST sequences
Homo sapiens	5 397 000
Mus musculus and *M. domesticus*	3 820 000
Rattus sp. (rat)	537 000
Triticum aestivum (wheat)	500 000
Ciona intestinalis	493 000
Gallus gallus (chicken)	442 000
Zea mays (maize)	359 000
Danio rerio (zebra fish)	357 000
Glycine max (soya)	339 000
Bos taurus (cattle)	320 000
Xenopus laevis (xenopus)	295 000
Drosophila melanogaster	262 000
Caenorhabditis elegans	216 000
Oriza sativa (rice)	203 000

5.4 The Transcriptome

In the same way that genomics represents the study of genomes, the transcriptome consists of establishing the expression profile of messenger RNAs on the grand scale. This approach was initially implemented with model species, but is currently being applied to man. Great outcomes are expected for both fundamental and pathological science.

5.4.1 Local analyses of transcription

A certain number of classical analyses allow the study of gene transcription. The technique of Northern blotting (Figure 1.7), for example, can be used to identify the tissues or stages wherein a gene is transcribed, and the size of the messenger. This approach allowed the demonstration of transcription of about 80 per cent of yeast genes.

In situ hybridization, carried out using a radioactive probe (RNA or DNA), may be used to study pathology in humans, using cells or tissue samples. In model species (yeast, drosophila or nematode), this approach can be applied to the whole organism. Large-scale *in situ* hybridization is, however, difficult to develop.

Transcriptional analysis may also be carried out by inserting a reporter gene – such as lacZ or GFP (green fluorescent protein) – downstream from a promoter that one wishes to study. LacZ encodes ß-galactosidase, and its expression is detected by the blue colour obtained in the presence of X-Gal. GFP is a protein containing a chromophore, which fluoresces under blue light (395 nm). Use of these reporters allows the evaluation of expression levels, and the identification of tissues in which the normal gene is expressed under the chosen promoter. Reporter genes have been used in various bacteria, *Drosophila* and *C. elegans* (Figure 5.3). This type of analysis is particularly interesting in the latter species, because the nematode is completely transparent: the specificity of expression can thus be observed in the living animal, after introduction of the reporter gene under control of the promoter.

5.4.2 Massive sequencing and the global vision of cellular physiology

Massive sequencing of cDNA from a given tissue allows one to establish the list of genes expressed in that tissue: the frequency obtained for each

Figure 5.3   Transcription analysis using reporter genes in *C. elegans*. (A) Study of the transcription of the *mec-9* gene, by fusion of its promoter to the gene encoding lacZ. (B) Study of the transcription of the *glr-1* gene, by fusion of its promoter to the gene encoding GFP. In both cases, the marked cells are those in which the reporter gene is transcribed, these are therefore the cells in which the gene is normally transcribed

sequence gives an indication of the level of expression of the gene concerned. In this way thousands of genes have become accessible to study. The analysis of this type of data produces a global view of the activity of cellular physiology. This type of analysis has been carried out by several groups, and some examples are given in Figure 5.4.

These data reflect certain peculiarities of the physiology of the organs concerned. One will note the importance of contractile proteins in the heart, of secreted and hormonal proteins in the pancreas and liver, of membrane proteins, the cytoskeleton and signal transduction in the brain. Metabolic proteins by contrast are uniformly found in all these tissues. This latter group seem to be involved in basal activities common to all cells.

This type of analysis has been carried out on a large number of tissues, and the number of genes associated with each major cell function has been estimated for many different tissues. This provides an estimate, over the whole genome, of the proportion of genes implicated in each major biological function of man (Table 5.2).

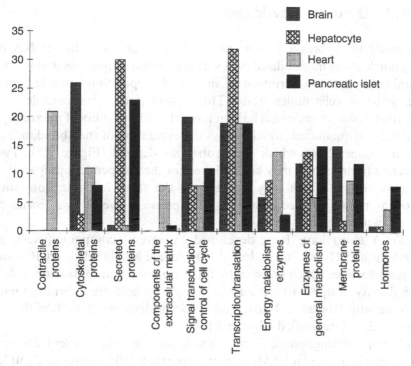

Figure 5.4 Quantitative distribution of messengers identified as a function of their biological role in different tissues

Table 5.2 Average distribution of messages identified as a function of their biological role over the whole organism

Functional type	Fraction of genes implicated
Cellular communication	12%
Transcription and maturation of RNA	6%
Protein synthesis	15%
Metabolism	16%
Replication and cellular division	4%
Cytoskeleton, cellular mobility	8%
Homeostasis, defense of the organism	12%
Unclassable	24%

5.4.3   Use of nucleic acid chips

To study the expression of a group of genes one may use cDNA or oligonucleotides immobilized on a surface: these are hybridized to probes obtained by retrotranscription of an mRNA population extracted from the tissue or cells under study. This system is thus comparable to a 'Northern blot' in reverse. The signal is detected by confocal microscopy and may be quantified, which allows the evaluation of the abundance of the messenger from which the probe was derived (Figure 5.5). Two different fluorophores may be employed for the independent preparation of two probes, which may be used in the same hybridization, thus allowing the simultaneous analysis of expression level in two tissues or under two distinct conditions.

cDNA chips have been developed by several companies, such as Syntheni and Hyseq. The selected cDNAs are amplified by PCR using oligonucleotides which are specific for them, and immobilized on a slide at a density of up to 10 000 samples per cm^2. These are hybridized with fluorescently labelled cDNAs, and the hybridization of each cDNA is measured and quantified (Figure 5.6).

Currently the highest-performing applications are oligonucleotide chips, developed in particular by Affymetrix. More than 250 000 oligonucleotides can be synthesized directly onto a surface (glass or silicon) of about 1.6 cm^2, by a procedure employing photolithography and DNA synthesis. These chips are hybridized to the RNA population emanating from a given tissue, after retrotranscription into labelled cDNA (Figure 5.7). The expression of each gene is evaluated using a score of distinct oligonucleotides, so as to optimize its recognition. Detection of hybridization signals allows the determination of which gene is expressed under this or that condition (the main problem with this approach is its cost, still very high).

mRNA expression profiles may be rapidly established in prokaryotes, where the majority of the genome is coding. Thus, 64 000 oligonucleotides specific for the genes of *Streptococcus pneumoniae* were synthesized and fixed, and their levels of expression in stationary and exponential growth phase were compared, using cDNAs synthesized from messengers extracted in these phases. This allowed the identification of genes active in the exponential phase: a significant number code for enzymes involved in biosynthesis of capsule components, of fatty acids, or in cellular division.

In yeast, 260 000 oligonucleotides corresponding to all the genes have been synthesized onto a 1.28 cm^2 microchip. These chips have allowed the identification of genes expressed in various mutants, under different

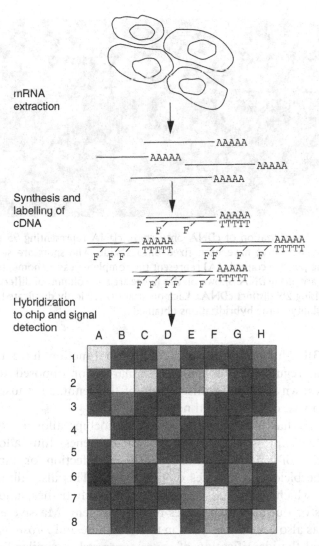

Figure 5.5 Use of chips carrying cDNA or oligonucleotides. Messenger RNAs are extracted from a tissue or cell culture, fluorescent complementary DNA is synthesized using labelled nucleotides (F), the messenger RNA is then eliminated, and the cDNA is hybridized to a chip containing an array of cDNA molecules or oligonucleotides. Detection allows the evaluation of the presence and abundance of each messenger RNA: for example the sequences matching B2 or E3 are abundant, those matching A1 or C2 are less frequent, and none match B1 or F1

culture conditions, or at different stages of growth [mitosis, diauxie (transition from fermentative to respiratory metabolism), oxidative stress, thermal shock, treatment with various drugs, and sporulation

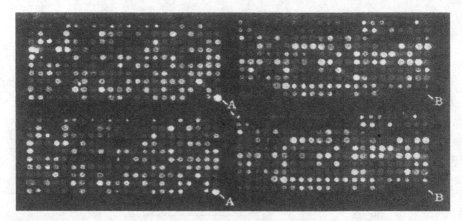

Figure 5.6 Hybridization of cDNA onto chips. cDNA representing 95 per cent of yeast coding sequences have been fixed onto a slide. The spots are separated by 380 μm. The positive controls (A) represent the complete yeast genome, the negative controls (B) are phage DNA. Shown on this figure are two columns of different matches each comprising 219 distinct cDNAs. Each quadrant is vertically duplicated, to control for the reliability of the hybridizations obtained

(Table 5.3)]. Numerous genes of unknown function have thus been recognized, regulated in a manner similar to or opposed to that of genes of known function; transcription of the genome is thus incorporated into a vast combinatorial network.

Affymetrix has also commercialized microchips allowing the evaluation of the expression of 8000 *A. thaliana* genes, thus allowing the identification of genes active during pathogen infection, or during treatment by herbicides, fungicides or insecticides. This also allows one to determine which genes are transcribed in which tissues, under which conditions or during which stages of development. Massive expression analysis has also been carried out in the nematode and *Drosophila*, which has allowed the identification of genes expressed in similar fashions in different tissues.

For humans, chips have been made carrying oligonucleotides complementary to about 8000 already-identified genes, and their expression in various tumour types (Figure 5.7) or genetic diseases such as rheumatoid arthritis has been studied. Chips have also been synthesized carrying oligonucleotides corresponding to genes predicted from the genomic sequence by various informatic programmes, for which no experimental data regarding transcription is yet available. Hybridization to various labelled cDNAs emanating from different human tissues (healthy or

Table 5.3 Yeast genes whose transcription is stimulated during different stages of meiotic sporulation

Stage	Form of spores	Meiotic stage	Number of genes stimulated/ principal regulatory factor/ promoter domain recognized	Role
Early induction (I)				DNA replication, sister chromatid cohesion, appearance of homologous chromosomes, recombination, polar dynamics
Early induction (I)			161/IME1/URS1 domain	
Intermediate induction			253/Ndt80/MSE domain	First meiotic division
Pre-late induction			61/Ndt80/MSE domain	Second meiotic division, formation of spores
Late induction			5/Ndt80/MSE domain	Formation of walls

cancerous) has allowed the demonstration that these genomic regions contain an exon in 55–65 per cent of cases. Absence of a positive result does not necessarily indicate a prediction error, it might mean that the message containing this sequence is not expressed in the particular tissues from which the labelled cDNAs were synthesized.

Finally – although here we move from the transcriptome to rejoin the genome – mutations in genes responsible for disease may also be identified using chips (Figure 5.8); for example, this has been carried out in the case of the CFTR gene (mutations in which cause cystic fibrosis), BRCA1 (breast cancer) and p53 (tumour suppressor gene).

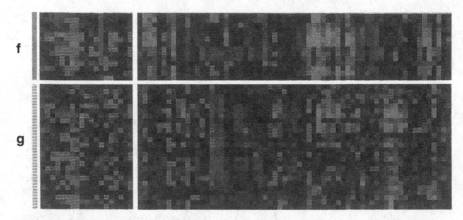

Figure 5.7 Hybridization on an oligonucleotide chip. Part of the oligonucleotide chip hybridization allowing the analysis of 8102 genes is shown. On the horizontal axis are oligonucleotides allowing the detection of transcription of some of these genes (group f: γ immunoglobulin, CSF (colony stimulating factor), NSF (neutrophil cytosolic factor), seven regions of the λ immunoglobulin chain, a class II MHC protein, μ immunoglobulin, EDR2 (early development regulator 2), MIP-I (max-interacting protein I); group g: homeobox-I, somatomedin C, KIP-2 (cyclin dependent kinase inhibitor P57), five different proteins interacting with fatty acids, two collagen receptors (CD36), a glutathione peroxidase, LIM protein, an alcohol dehydrogenase, a glycerol-phosphate dehydrogenase, a protein recognizing retinol, a lipase, α7 integrin, four channels of the aquaporin type, AI apolipoprotein, a cytokinin, and two receptors for endothelin). Each oligonucleotide occupies a 24×24 μm surface. These chips have been hybridized to cDNA originating from 84 breast-cancer patients, before or after treatment with doxorubin (vertical axis)

Sequencing a given region is much simpler and cheaper if the sequence has already been determined previously: research into sequence variation by comparison with an already-known sequence is thus rapidly carried out. The problem with this approach is the presence of repeated sequences, which prevent the identification of certain regions. However, this strategy has been used in the identification of mutations in the human mitochondrial genome (Figure 5.9), which may cause genetic diseases (this 16 kb genome contains no repeated sequences).

5.4.4   The SAGE technique

The SAGE technique (serial analysis of gene expression) consists of identifying the collection of genes transcribed in a given tissue or devel-

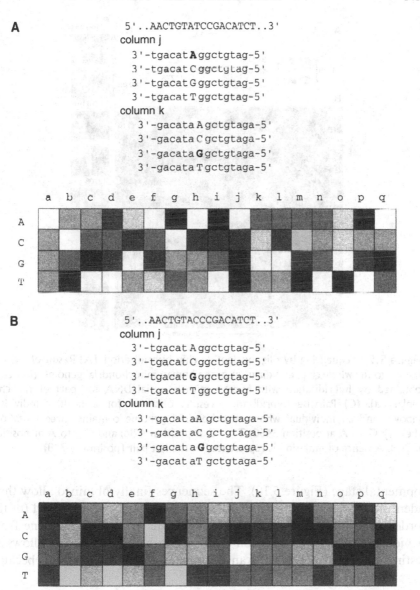

Figure 5.8 Mutation detection on an oligonucleotide chip. The chips represented bear 68 quadrants, each carrying 15-base oligonucleotides. Their sequence is complementary to the unmutated sequence, but for the seventh nucleotide, the four possible nucleotides are represented in four distinct quadrants (in a column): the oligonucleotide contents of columns J and K are represented for four quadrants. In (A) the signal obtained for hybridization in the 10th column/first quadrant shows that the hybridizing DNA has the sequence 5'-ACTGTATCCGACATC-3', complementary to the oligonucleotide sequence 5'-tgacatAggctgtac-3'. The other three quadrants of the

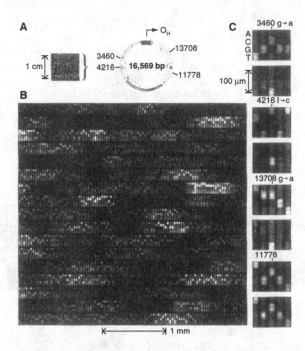

Figure 5.9   Sequencing by oligonucleotide chip hybridization. (A) Result of hybridization to the whole chip, and diagram of the human mitochondrial genome. (B) Result obtained by hybridization with human mitochondrial DNA for part of the chip (enlarged). (C) Pairwise hybridization results obtained for a healthy individual (above) and an individual whose mitochondrial genome contains three mutations (below): G → A at position 3460; T → C at position 4216; and G into A at position 13708. Absence of mutation is represented in the last pair (position 11 778)

opmental stage (Figure 5.10). The sequences finally obtained allow their identification: for each gene, the size of the sequence is very short (of the order of a dozen nucleotides), but adequate to identify the gene from which it derives, by comparison with the databases. SAGE also allows an estimate of the frequency of transcription of each identified gene, because

---

same column are not complementary (seventh base of each oligonucleotide shown in bold), and do not hybridize. Similarly, the signal obtained in the 11th column shows that the hybridizing DNA has the sequence 5'-ACTGTATCCGACATC-3'. This sequence is therefore not mutated. In (B) a single positive hybridization signal has been obtained (10th column, third line), showing that only the oligonucleotide 5'-tgacatGggctgtag-3' has hybridized, and that the DNA sequence is 5'-AACTGTACCCGACA-3'. The other regions have not hybridized, because of non-complementarity of at least one base (shown in bold). This sequence is therefore mutant

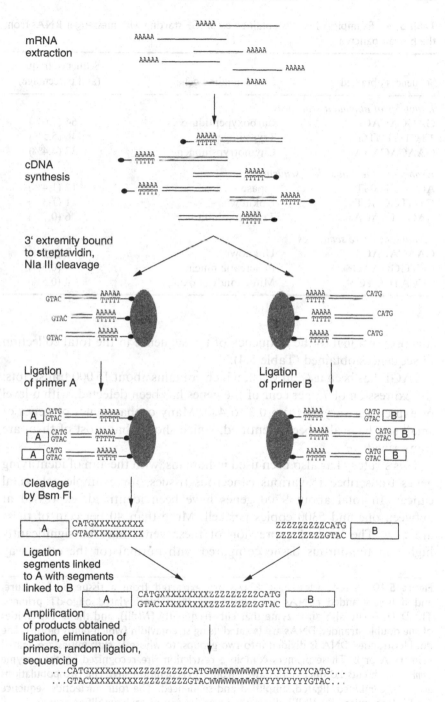

mRNA
extraction

cDNA
synthesis

3' extremity bound
to streptavidin,
NIa III cleavage

Ligation
of primer A

Ligation
of primer B

Cleavage
by Bsm FI

Ligation
segments linked
to A with segments
linked to B

Amplification
of products obtained
ligation, elimination of
primers, random ligation,
sequencing

Table 5.4   Examples of results obtained by SAGE starting with messenger RNAs from the human pancreas

Sequences obtained	Corresponding gene	Sequence frequency (and percentage)
Examples of abundant sequences		
GAGCACACC	Carboxypeptidase	64 (7.6%)
TTCTGTGTG	Trypsinogen	46 (5.5%)
GAACACAAA	Chymotrypsinogen	37 (4.4%)
Examples of less abundant sequences		
AGCCTTGGT	Lipase	12 (1.4%)
GTGTGCGCT	Unknown	11 (1.3%)
AAGGTAACA	Trypsin inhibitor	6 (0.7%)
Examples of rare sequences		
GAACACACA	Unknown	4 (0.5%)
CCTGGGAAG	Pancreatic mucin	4 (0.5%)
CCCATCGTC	Mitochondrial oxidase	4 (0.5%)

it is proportional to the frequency of the sequence in the total collection of sequences obtained (Table 5.4).

SAGE has been used in yeast, which contains about 15 000 transcripts: the expression of 75 per cent of the genes has been detected, with a level of expression varying from 0.3 to 400. Many orphan genes or genes of small size have also been identified, which shows that most of them are transcribed.

This strategy has also been used in humans, with the aim of identifying genes transcribed in various cancerous tissues, for example colorectal cancer. In total about 9700 genes have been identified, expressed at between one and 5300 copies per cell. More than 40 per cent of these are new. The level of expression of these genes may be significantly higher in tumourous tissue compared with normal (or the other way

Figure 5.10   SAGE. Messenger RNAs are extracted from a tissue or a culture, and double-stranded cDNAs are synthesized from biotinylated oligo-dT primers. The DNA is cut with an enzyme that cuts frequently (*Nla*II), and the 3′ extremities of the double-stranded DNAs are isolated using streptavidin (which binds biotin). The double-stranded DNA is divided into two groups, to whose 5′ extremities are ligated primers A or B. These primers contain a restriction site recognized by the enzyme *Bsm*FI, which cuts 20 nucleotides away from its recognition site. The two populations are then combined, ligated, amplified and sequenced. The four nucleotide sequence CATG (recognized by *Nla*II) allows the identification of each amplified region

round). Amongst the 548 genes expressed at different levels, 337 were already known. Genes expressed at higher level in the normal colon include for example genes involved in physiology or tissue architecture, e.g. cytokeratin, carbonic anhydrase, guanyline. Many genes expressed at higher level in the tumourous colon are linked to growth, e.g. ribosomal proteins, elongation factors, enzymes of glycolysis, IGFII (insulin-like growth factor II), filamin (linked with actin). The fraction of genes showing very different transcription levels in normal compared with cancerous tissues is nevertheless astonishingly small: of the order of 1.5 per cent.

Similar approaches have been used for the identification of human genes whose expression is modified in other diseases, such as cancer of the brain, lung, gastro-intestinal tract; or atherosclerosis, infection by cytomegalovirus or HIV-1. They have also been employed in various studies on the mouse, *Arabidopsis* and rice.

5.5 Conclusion

The cDNA sequencing programmes have thus demonstrated their possible application in humans or other organisms, in the physiological domain (specifically the studies of tissues, cells or stages of development), and in pathology. The functional analysis of a formidable collection of novel genes remains to be done; every unknown sequence is a research project in itself. These data are also widely used for the identification of coding sequences within the global human genome sequence, and to this end, complete cDNA sequencing is currently being developed.

6 The Proteome

The proteome is the collection of proteins being expressed by a cell or tissue at a given moment. The links between genome, transcriptome and proteome are complex, because several distinct messages may be made from the same gene, and several proteins may be made from the same messenger.

The salient features of proteome analysis by comparison with those of the genome or transcriptome lie in the following characteristics: (i) the same genome may give rise to many different proteomes: in *M. genitalium* one finds 24 per cent more distinct proteins than distinct genes, in humans this ratio can be of the order of 200–300 per cent; (ii) the level of expression of a protein is not predictable from the level of expression of its messenger (the abundance of a protein can be up to 50 times different from that of its messenger); (iii) messengers may show considerable diversity, because of the use of distinct promoters, or alternative splicing; (iv) proteins may be translated from different initiation codons, and are frequently post-translationally modified (by glycosylation, phosphorylation, prenylation, acetylation, ubiquitination, deamination...) – the same gene may thus give rise to a large number of distinct proteins; (v) it is very difficult to predict, starting from the analysis of DNA or cDNA sequences, under what conditions a gene is transcribed and translated, in which stage of development, or in which tissue; and (vi) the number of different proteins expressed in tissues is of the order of several thousand (10 000–15 000 on average, 30 000 in nerve cells).

Proteomics is thus of great importance, allowing for example the verification of whether this or that predicted coding sequence is actually translated and present amongst the proteins of an organism. Nevertheless it presents a certain number of difficulties: (i) no enzyme allows the amplification of proteins (as PCR does for nucleic acids); (ii) the presence

Genome, Transcriptome and Proteome Analysis by Alain Bernot
© 2004 John Wiley & Sons, Ltd   ISBN 0 470 84954 1 (HB)   ISBN 0 470 84955 X (pbk)

of this or that protein after extraction may be more or less efficient depending on the type of solubilization performed, and there is no extraction medium in which all proteins are soluble; (iii) feebly expressed proteins are difficult to detect; (iv) proteomic analysis gives little chance of detecting point mutations, insertions or small deletions; and (v) the number of proteins identified may be enormous, but their function is often unknown (for only 8000 human proteins is any function 'known').

6.1 Basic Techniques

6.1.1 Electrophoresis

Protein electrophoresis allows one to separate the collection of proteins present in a cellular extract, using an electric field. In the SDS–PAGE method (SDS: sodium dodecyl sulphate; PAGE: polyacrylamide gel electrophoresis), this separation is carried out in a polyacrylamide gel, in the presence of SDS, an anionic denaturing agent which abolishes non-covalent bonds, and mercaptoethanol or dithiothreitol, which reduce disulphide bridges. The speed of separation depends on the mass of each protein, small ones migrating more rapidly than large ones (Figure 6.1).

The gel may subsequently be stained using Coomassie blue (or silver nitrate, which is more sensitive), and the separation generally allows one to distinguish between two proteins whose respective masses differ by 2 per cent. Another approach uses methionine ^{35}S, which is introduced into the culture medium before protein extraction, thus marking those proteins newly produced by living cells: these may be detected by auto-radiography. Figure 6.1 shows the variation in migration of haemoglobin from individuals affected by sickle-cell anaemia compared with the normal version: this characterization, shown by Ingram in 1958, was the first evidence for an aminoacid substitution (glutamate for valine at position 6) that causes a human genetic disease.

Another approach, the Western blot, consists of transferring the result of protein electrophoresis to a nitrocellulose membrane. The membrane may subsequently be exposed to an antibody specific for the protein of interest: if this antibody is radioactively labelled, the presence and local-ization of the protein may be obtained by autoradiography.

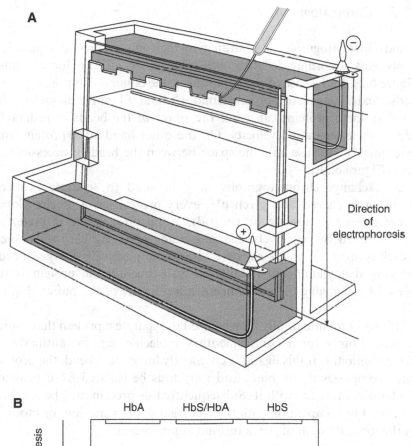

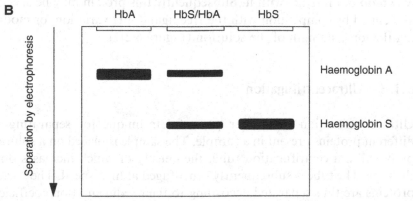

Figure 6.1 Protein electrophoresis. (A) Migration under the influence of an electric field in an acrylamide gel. (B) Examples of the migration of normal globins (haemoglobin A) and mutant globin (haemoglobin S), the latter being responsible for sickle-cell anaemia in homozygotes

6.1.2   Chromatography

Liquid chromatography is a strategy which allows precise separation of proteins according to their size, charge or affinity for a ligand (Figure 6.2). In the case of size separation, the column comprises porous beads (dextran, agarose, polyacrylamide), about 0.1 mm in diameter. The fact that small proteins can enter the pores of the beads considerably reduces their separation velocity. On the other hand large proteins migrate quickly, because only the space between the beads is accessible to them (Figure 6.2B).

Ion exchange chromatography may be used to separate proteins according to charge: at a given pH, every protein has a precise global charge (negative, positive or neutral). A positive charge is preferentially retained by negatively charged beads (carrying carboxyls, such as carboxymethyl), and a negative charge by positively charged beads (carrying diethylaminoethyl, DEAE). Subsequently the protein is recovered by changing the salt concentration or pH of the buffer (Figure 6.2C).

Affinity chromatography may be used to separate a protein that shows elevated affinity for a complementary molecule, e.g. an antibody or enzyme inhibitor. If this ligand is covalently linked to a bead, the protein which recognizes it will bind, and may thus be isolated from proteins which do not interact with it. Subsequently this protein may be specifically eluted by competition with the free ligand, pH variation, or modifying the ionic strength of the solution (Figure 6.2D).

6.1.3   Ultracentrifugation

Ultracentrifugation is another powerful technique for separating the different proteins present in a sample. The sample is placed on a solution, prepared in a centrifugation tube, the density of which increases down the tube. The tube is subsequently centrifuged at high speed. The various proteins are thus separated according to their sedimentation coefficient, which increases with size (sedimentation coefficients are expressed in Svedberg units, S). During centrifugation, heavier proteins sediment more rapidly than lighter proteins. Centrifugation is generally stopped once the heaviest proteins reach the bottom of the tube. The different proteins thus separated form bands, which are individually recovered by collecting successive fractions from the tube (Figure 6.3).

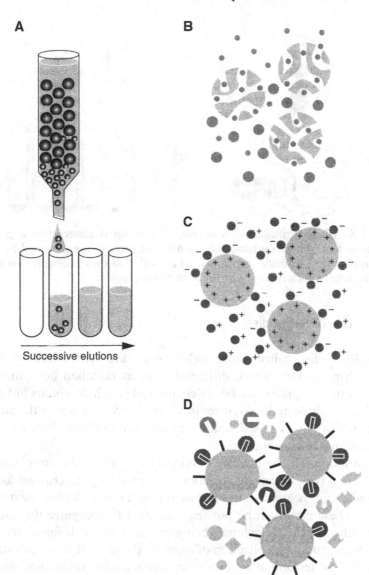

Figure 6.2 Protein chromatography. (A) Chromatography allows protein separation by the use of beads able to retain a particular type. (B) Chromatography by filtration or steric exclusion: porous beads retain the small proteins but not the large ones, which thereby elute more quickly. (C) Ion-exchange chromatography: in the case shown, positively charged beads retain negatively charged proteins. Positively charged proteins are eluted. (D) Affinity chromatography: proteins having an affinity for the substrate which is covalently linked to the bead are retained, the others are eluted. The protein is recovered when free substrate is introduced into the column

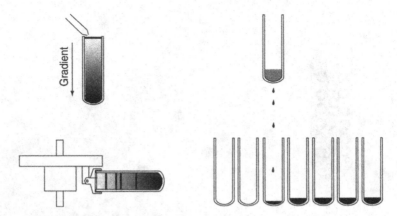

Figure 6.3 Ultracentrifugation. A mixture of proteins is placed onto a gradient prepared in a centrifugation tube, and the different proteins are separated by ultra-centrifugation (high-velocity centrifugation) according to their sedimentation coefficient. Preparation is carried out as a function of this gradient

6.1.4   Use of antibodies

Antibodies – also called immunoglobulins – are synthesized by verte-brate B lymphocytes, which differentiate in mammalian bone marrow. An antibody comprises two heavy chains and two light chains linked by disulphide bridges, and recognizes in a highly specific manner the antigen against which it was raised, thanks to two binding sites located at the proteinaceous N-terminals.

An antibody may be directly obtained by injecting the antigen into a rabbit: after two successive injections separated by several weeks, the prepared serum is greatly enriched in antibodies specific for the injected antigen. These may then be purifed, and used to recognize the antigen. Such antibodies are polyclonal, being synthesized by different lympho-cytes which produce a collection of immunoglobulins that are not strictly identical. The technique of producing monoclonal antibodies was de-veloped by Köhler and Milstein in 1975 (Figure 6.4): the first stage consists of injecting a mouse with the antigen against which one wishes to raise an antibody. A large number of B lymphocytes is thus obtained, and these are fused to a myeloma, thus becoming immortalized: these clones retain their specific antibody production. One may then determine which clones produce antibody specific for the injected antigen.

The uses of antibodies are very diverse: for example they may be fixed to a polyvinyl choride support, which allows estimation of the concen-

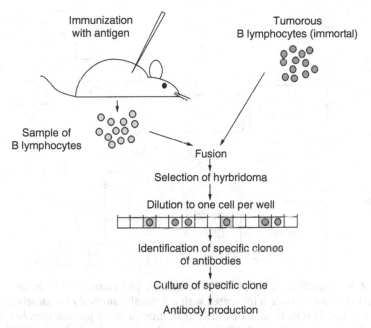

Immunization
with antigen

Tumorous
B lymphocytes (immortal)

Sample of
B lymphocytes

Fusion

Selection of hyrbridoma

Dilution to one cell per well

Identification of specific clones
of antibodies

Culture of specific clone

Antibody production

Figure 6.4 Production of monoclonal antibodies in the mouse. After injecting an antigen into a mouse, the B lymphocytes are prepared, and fused with tumour cells. The hybridomas are subsequently selected, diluted down to one cell in wells, and those which secrete the specific antibody are indentified

tration of the cognate antigen in a given solution. Fixation to a support also allows the identification of protein 'preys', which interact with the protein 'bait' recognized by the antibody (Figure 6.5A). An antibody may also be used as a marker, if it is radioactively or fluorescently labelled. Thus, in a Western blot, after protein migration and transfer to a membrane, one may detect a protein using the antibody which is specific for it. This type of identification may also be carried out on tissue or cell sections, which allows one to establish the cells or parts of cells within which the protein of interest is present.

Specific antibodies also allow expression library screening: this approach may be used where a protein has been characterized only by antibodies, and one is seeking to identify the gene which encodes it. mRNAs are first purified from the tissue which expresses this protein, reverse transcribed into cDNA, and cloned into an expression vector (for example λgt11 or λ ZAP II). The library thus obtained is spread onto Petri dishes, the expression of the cloned genes is induced, and the proteins are transferred onto nitrocellulose filters. These filters are

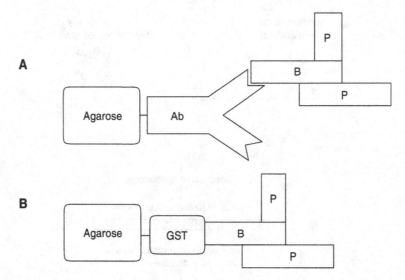

Figure 6.5 Identification of protein interactions. (A) Immunopurification of a bait protein (B). Protein lysate is incubated with a B specific antibody covalently linked to agarose beads. This method allows the purification of the B protein together with its interacting partners P (preys). (B) Fusion to glutathione S-transferase (GST) to a B protein: the hybrid protein GST-B allows the identification of P proteins interacting with B. The complex can be isolated using agarose beads

incubated with specific antibodies, and a positive signal allows the identification of the clone containing the cDNA that encodes the desired protein (Figure 6.6).

Use of antibodies thus has many applications, but this tool is not applicable to the analysis of the whole proteome in a eukaryote: one comes up against the practical problem of obtaining antibodies against all proteins (more than 19 100 in the nematode, without counting alternative splicing and post-translational modifications).

6.1.5   Study of protein interactions

We have seen above how proteins interacting with a given protein may be identified by the use of specific antibodies (Figure 6.5A). The analysis of protein interactions is fundamental, because the majority of proteins interact with other proteins, forming complexes.

Another analysis of protein interactions is possible if the gene encoding the 'bait' protein is available. This gene is cloned as a fusion with the gene encoding glutathione S-transferase (GST, conserving the reading

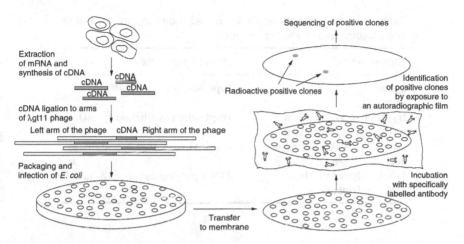

Figure 6.6 Expression screening. The mRNAs are retro-transcribed, and inserted into a phage library (here λgt11). The library is grown in *E. coli*, and the expressed proteins are transferred to a membrane. Clones expressing the desired protein are identified using a specific antibody. They are subsequently sequenced

frame), and the resulting hybrid protein is linked to agarose. After incubation with a collection of proteins, the 'prey' proteins recognizing this protein may be identified (Figure 6.5B).

Large-scale biochemical analysis of this type has been carried out for example in yeast. A collection of 6100 clones was produced, in each of which a yeast gene was associated with that encoding GST within a plasmid. These clones were analysed in 64 pools, each comprising 96 clones. Having identified a pool shown by biochemical testing to exhibit a given biochemical activity, the clone responsible could be isolated by analysing individual clones of the pool, separated into eight rows and 12 columns. After isolating the GST-ORF, the activity of the protein could be analysed, and the responsible gene identified. In this way genes were isolated that are involved in tRNA ligation, the 2'-phosphotransferase reaction, and methyl-transferase activity.

6.1.6 Informatics

The function of a protein produced by a given gene may be proposed based on its homology to another gene of known function. This approach is, however, not totally reliable, and experimental confirmation has to be carried out to confirm it.

Table 6.1 Independent genes in *E. coli,* homologous to a single homologous gene in another species

Escherichia coli	Homologous gene
gyrase A	topoisomerase II (yeast)
gyrase B	
tryptophan synthetase α	tryptophan synthetase (yeast)
tryptophan synthetase β	
acetate CoA-transferase α	succinyl CoA-transferase (man)
acetate CoA-transferase β	
DNA polymerase III α	DNA polymerase III α (*B. subtilis*)
DNA polymerase ε	

To a certain extent protein interactions may be predicted by comparing genome organization between different species. Comparing the genome of *E. coli* with those of other organisms has thus allowed the identification of a certain number of separate genes that are all similar to one gene in another species: it is probable that such genes encode proteins which interact (Table 6.1).

Protein domains may be recognized to a certain extent using informatic tools. Currently, about 2000 domains have been identified, and amongst the entirely sequenced genomes they may be present in 40–50 per cent of proteins. In prokaryotes, the most abundant domain is the ABC transporter; in eukaryotes, the protein kinase domain.

6.2 Transgenesis

Transgenesis consists of introducing a supplementary gene into an organism. If the transgene integrates into the host genome the transgenesis is stable; if it does not integrate the transgenesis is transient. In the first case, the gene is transmitted to all the descendants of the host. Expression may be ubiquitous or tissue-specific, according to the promoter cloned upstream of the gene. Several techniques for DNA transfer exist, and the diversity of species susceptible to transgenesis is vast: mouse, rat, rabbit, pig, sheep, cattle, chicken, amphibians, fish, insects, nematodes and plants. Transgenesis also allows one to breach the barriers between species, and to explore the genetic diversity identified throughout life.

6.2.1 Mouse transgenics

Mouse transgenics are obtained by pronuclear injection of DNA into the zygote: super-ovulation is first induced in the female, and the zygotes are prepared. Some 200–300 copies of a gene linked to a promoter are then introduced into one of the two pronuclei prior to fusion, and the surviving eggs are implanted into pseudo-pregnant females. Practically all types of DNA may be employed, but genomic fragments are more generally used than cDNA for transgenic expression (often because of enhancers within the genomic fragment). Integration is random, equally frequent in autosomes and sex chromosomes, and the animals thus obtained are transgenic in 10–30 per cent of cases (Figure 6.7). Recombination is heterologous, taking place between non-homologous regions of DNA. In general, incorporation occurs very early in development, so the animal obtained is not a mosaic of transgenic and non-transgenic cells. However, incorporation may occur later, after several mitoses have already taken place. Some cells are then transgenic, others not: in this case the animal is a chimaera.

There is an important positional effect of the site of integration: the presence of nearby regulatory elements, the availability of the region for transcription and the presence of transcriptional regulatory factors may play a major role in the level of expression obtained. Several transgenic animals are thus generally produced to overcome such problems.

The advantages of pronuclear injection are the relatively high frequency of obtaining non-chimaeric transgenics, the slight limitation on the size inserted and its stability. The disadvantages are the possible influences of integration site, the resultant insertional mutagenesis of the host genome and the absence of integration into the host germline in a chimaera.

Retroviruses may also be used in transgenesis. The fragment to be inserted is first integrated into the virus genome, from which the genes normally implicated in its replication and propagation have been removed. The virus is then used to infect embryonic cells at the four to 16 cell stage (only mitotic cells can be infected with a retrovirus). As before, integration into the genome is random. The disadvantages of this approach are the small number of integrations, the limited size of fragments clonable in a retrovirus (of the order of 15 kb), the frequent occurence of chimaeric embryos, the possible influences of retroviral sequences with transgene expression and, as before, the effects of integration site and insertional mutagenesis.

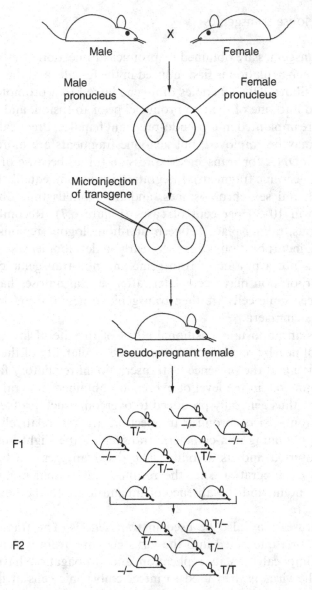

Figure 6.7 Transgenesis in the mouse. The gene of interest is injected into one of the pronuclei of an egg prepared from a fertilised female, and the egg is then reimplanted into a pseudo-pregnant female. Individuals heterozygous for the transgene may be identified in the Fl. Crossing two Fl heterozygotes gives an F2 generation amongst which one-quarter of the mice will be heterozygotes

Another approach consists of using embryonic cells (ES, embryonic stem cells). These cells are totipotent, originating from cell cultures prepared from the embryo before separation of the somatic and germ lines. The gene of interest is injected into these cells, and the incorporation of the transgene is checked by PCR. Positive cells are introduced into blastocysts, and, if they are incorporated into the germ line, all following generations will be transgenic.

Integration of the transgene in the mouse is checked by extraction and analysis of genomic DNA. Transcription of the transgene may be evaluated by northern blot, inverse PCR, *in situ* hybridization, or by study of encoded protein expression using specific antibodies.

6.2.2 Examples of mouse transgenics

One of the first studies of regulation of gene expression using transgenic mice was carried out by Brinster and Palmiter (1982), who studied the expression of growth hormone under the control of larger or smaller fragments of the regulatory region of elastase (an enzyme of intestinal digestion). In this way they studied the expression of growth hormone in pancreatic cells (in non-transgenic mice, elastase is only produced in pancreatic cells, and growth hormone by the cells of the pituitary gland). These workers also introduced a gene encoding human growth hormone into the mouse, under the control of the metallothionin promoter and enhancer. In some cases, the quantity of hormone secreted was 800–1600 times greater than normal, with the spectacular result of a mouse twice as big as normal.

The number of transgenic mice obtained to date is very large, these having been produced with the aim of studying the function or expression of various genes from the same or different species. Several private companies whose aim is the production of transgenic mice have been formed: these include Transgène (France), PPL Therapeutics, Genzyme Transgenics Corporation, Grace and Somatogen (USA).

Transgenesis sometimes leads to a completely unexpected phenotype, resulting from insertion of the transgene within a gene, thus leading to insertional mutagenesis (mutation of a host gene). Such events may be used to identify the role of the protein whose expression is thus compromised. Sometimes this has led to the identification of a new gene; for example, a transgenic mouse led to the identification of the *inv* gene, encoding a protein implicated in left–right asymmetry in the mouse. A similar mutation (autosomal and recessive) has been observed in the human, leading to Kartagener's syndrome.

6.2.3 Trangenesis in other species

Numerous mammals have been the object of transgenesis, either for reasons of productivity (e.g. transgenic cattle have been produced with the aim of obtaining milk rich in casein and lactose), or with the object of producing therapeutic proteins usable in humans (Table 6.3).

In *Drosophila*, transgenesis is carried out using transposon P. This element can integrate stably into the genome, and express a transgene in a regulated manner. In the nematode, the DNA is microinjected into the gonadal syncytium.

Transgenesis of dicotyledonous plants is possible using the Ti (tumor inducing) plasmid derived from the bacterium *A. tumefaciens*, responsible for crown gall. These plasmids, having been modified (in particular by elimination of the regions responsible for tumourigenesis), are excellent vectors allowing the integration of exogenous DNA into the genome (Figure 6.8). Other systems allow one to obtain transgenic plants: biolistics (using a DNA gun to bombard cells with particles coated with DNA), liposomes (phospholipid membranes surrounding a molecule of DNA), and electroporation (brief electric shock allowing the breaching of membranes so that the DNA may enter the cell). Plant transgenics has been particularly used to create plants resistant to viruses, insects or herbicides.

Table 6.2 Pharmaceutical proteins produced by plant transgenesis

Transgenic plant	Protein produced (intended use)
Tobacco	Protein C (anticoagulant), erythropoietin (anaemia), EGF (division – cellular repair), haemoglobin, glucocerebrosidase (Gaucher's disease)
Rice	α1 antitrypsin (haemorragia, cystic fibrosis), α-interferon (liver treatment)
Tobacco, tomato	Albumin (cirrhosis), angiotensin inhibitor (hypertension)
Maize	Aprotinin (trypsin inhibitor)
Nicotiana bethamiana	α-Trichosanthin (inhibition of HIV replication)

Table 6.3 Various transgenic species created in the interests of furthering productivity of pharmaceutically interesting proteins

Transgenic species	Transgene	Gestation time/age of lactation (in months)	Quantity of recombinant protein (kg/year)
Rabbit	α1 antitrypsin (haemorragia, cystic fibrosis) Interleukin 2 (immune response) Plasminogen activator (treatment of infarctus) Superoxide dismutase (detoxifying enzyme) Calcitonin (calcium regulator) Erythropoietin (stimulation of erythropoiesis) Glucagon (glucose regulation)	1/7	0.02
Pig	Protein C (anticoagulant) Factor VIII (coagulation) Haemoglobin (treatment of thalassemias)	4/16	1.5
Sheep	Factor VIII Factor IX (coagulation) Fibrinogen (coagulation)	5/18	2.5
Goat	Plasminogen activator (treatment of infarctus) Interferons (liver treatment, multiple sclerosis) Antithrombin III (anticoagulant) Lactoferrin (fixer of iron) Antihaemophilic factor α1 antitrypsin (treatment of respiratory emphysema) Monoclonal antibody RB-96 (anticancer therapy) Growth hormone	5/18	4
Cow	α-lactalbumin (premature babies) Insulin (diabetes) Erythropoietin (anaemia, chronic kidney disease) Interferons	9/33	40–80
Chicken	Lysozyme (bactericide) Growth hormone Insulin	1/—	0.25

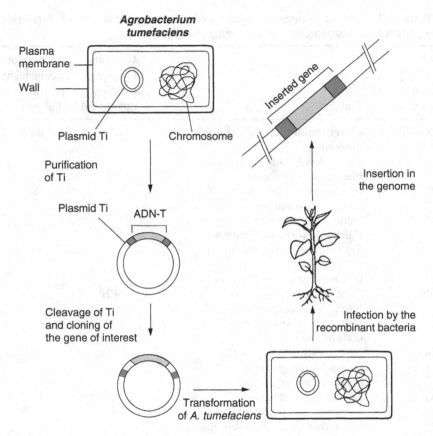

Figure 6.8 Transgenesis in plants. The gene of interest is cloned into the Ti plasmid, which is reintroduced into *Agrobacterium tumefaciens*, and used to infect plants. This gene is thus transferred into the genome of the infected plant

6.3 Mutagenesis

Mutagenesis consists of the production of individuals in which a gene has been mutated, which may allow comprehension of the function of the encoded protein, and to some extent the extrapolation of the results to humans (of particular interest in genes responsible for diseases).

Two approaches have been developed for the production of mutations: random mutagenesis, and site-directed mutagenesis. In the first case, the mutations are introduced randomly into the genome, whilst in the second, a point mutation is specifically introduced into a pre-selected gene. Once the mutation has been made, the mutants obtained are tested as part of the study of the function of the encoded protein.

6.3.1 Directed mutagenesis

In bacteria, site-directed mutagenesis is relatively easily performed because of the low complexity of the genome, and its ease of access. Mutant production may be carried out using a transposon: this is introduced *in vitro* in a part of the chromosome which includes the gene one wishes to mutate, and this genomic region is subsequently inserted into the bacterial chromosome by homologous recombination (which eliminates the original gene). Because of the insertion of the transposon into the ORF, the gene is no longer active. Phenotypic consequences of the mutation may be investigated as functions of culture condition: on more or less rich media, at different temperatures, in the presence of various inhibitors, or with respect to the abilities to grow or sporulate.

Yeast also lends itself quite easily to directed mutagenesis, because the complete genome sequence is known, and in this species homologous recombination is very effective (Figure 6.9). A gene may be replaced by a mutant allele (or a resistance gene, such as kanamycin), or even a barcode (short nucleotide sequence), which allows one to identify which gene is mutated within the population. Because yeast can maintain itself either in a haploid or a diploid state, replacement is first carried out in the diploid, and the strain obtained is heterozygous for the mutated gene. Subsequently the haploid or homozygous diploid line is produced, and the consequences of the mutation may then be analysed. If the mutation is lethal, half the expected haploids are absent (only those with the non-mutated gene survive), but the function of such genes may be studied using regulatable promoters or conditional mutants.

Evaluation of the functional role of a protein is established by the phenotypic modifications which follow mutation of the gene which encodes it. This may be carried out by culturing the mutants in different media – rich medium or poor, high in lactate, growth at elevated temperatures (30 or 36.5°C, instead of 25°C), high salt concentration – or over many generations.

A collection of mutants covering the whole yeast genome has been obtained. Each one contains a single mutation of a single gene (associated with a bar code), and the complete collection represents mutations in all the genes. This allows the development of a global approach to proteomics: a growth deficiency of a mutant, even a slight one, will result after several generations in domination by individuals not affected by the culture conditions. If a large population of mutants is cultured under particular conditions (e.g. minimal medium, exposure to UV, high salt

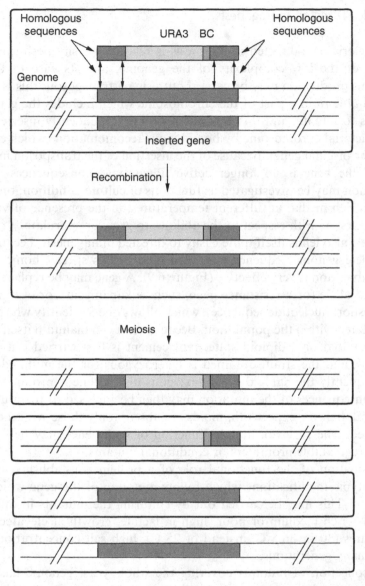

Figure 6.9 Directed mutagenesis in yeast. A gene may be replaced in a diploid by a resistance gene (URA3) associated with a 'bar-code' (BC). This region is flanked by sequences homologous to those which in yeast flank the gene one wishes to neutralize. After recombination, one of the two alleles is thus eliminated, and replaced by URA/BC. The haploid strain, in which the gene has been neutralized, may be isolated after meiosis (unless the mutation is lethal, in which case one only obtains haploid strains containing the unmutated gene)

concentration, absence of a particular metabolite, diauxie), the genes indispensable for growth under these conditions may be identified thanks to the bar code.

The analysis of this collection of mutants shows that a growth defect is observed in a third of cases, about 20 per cent of proteins are indispensible, and only half of genetic mutations cause no detectable physiological alteration. A large number of proteins is thus involved in growth. This underlines once again our lack of knowledge of the physiology of even a simple organism like yeast. Additionally, the identification of indispensible proteins (especially those whose sequence shows no similarity to any known human sequence) may be very useful in the development of antifungals.

In *C. elegans*, the equivalent of gene mutation may be obtained by the strategy of RNA interference (RNA-i). This consists of introducing two RNAs, one similar to that encoded by the wild-type gene one wishes to study, the other complementary to it. Two specific enzymes then recognize the double-stranded RNA, and cleave it. The host gene, though transcribed, is no longer translated (this enzymatic system probably evolved as a mechanism to protect the genome from transposon or retrovirus invasion). Introduction of RNA may be easily done in the nematode; for example by feeding it bacteria containing genes which encode the RNA, or injection of double-strand RNA into the zygote. Of the order of 60 000 double-strands are injected, and even though after development only a few double-strands per nucleus remain, the interference obtained is very effective and specific for the gene tested.

These approaches have been used to study about 4600 proteins, and in 472 cases phenotypic modifications have been identified. The majority (60 per cent) manifest themselves by embryonic death, and have been identified as proteins involved, for example, in chromosome condensation, chromosome separation, signal transduction and neuromuscular defects. A large fraction of the proteins thus identified has homologues in other multicellular organisms (45 per cent), whereas globally only 13 per cent show this kind of conservation. This suggests that the genes encoding them are significantly conserved in metazoa.

6.3.2 Directed mutagenesis in mice (*knock–out*)

To perform directed mutagenesis in the mouse, *knock-out* is performed on ES cells. A mutated version of the gene being studied, flanked by

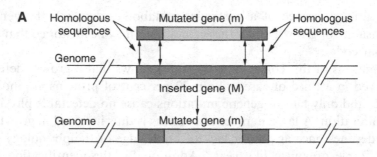

A

Homologous sequences — Mutated gene (m) — Homologous sequences

Genome

Inserted gene (M)

Genome — Mutated gene (m)

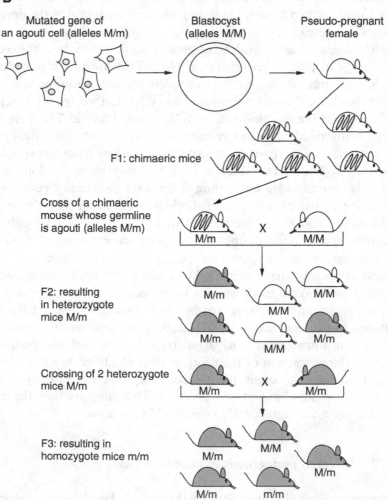

B

Mutated gene of an agouti cell (alleles M/m)

Blastocyst (alleles M/M)

Pseudo-pregnant female

F1: chimaeric mice

Cross of a chimaeric mouse whose germline is agouti (alleles M/m)

M/m X M/M

F2: resulting in heterozygote mice M/m

M/m   M/M   M/M   M/m   M/M   M/m   M/M

Crossing of 2 heterozygote mice M/m

M/m X M/m

F3: resulting in homozygote mice m/m

M/m   M/M   M/m   M/m   m/m

wild-type genomic sequences, is introduced into ES cells (frequently associated with a gene for antibiotic resistance, which allows a preliminary selection for transgenic cells). The *agouti* mouse strain may be used, because it is easy to identify by colour.

Homologous recombination between the sequences flanking the mutated gene and those flanking the host gene results in replacement of the target gene by the mutant gene (Figure 6.10). The embryonic cells in which the genome is modified are identified by PCR, and introduced into a blastocyst. From the resulting generation one selects those mice whose germline is *agouti*: their cells will thus be heterozygous for the gene. Uniformly heterozygous mice are obtained by a subsequent cross, and entirely homozygous mice may be obtained by crossing these heterozygotes.

6.3.3 Examples of knock-out in mice

The use of knock-out has allowed the verification of a number of hypotheses; this is particularly the case for mutations encoding proteins involved in the immune response (genes encoding immunoglobulins or T lymphocyte receptors). The knock-out of the gene encoding cadherin E, an adhesion molecule particularly involved in epithelial function, confirmed the vital role of this protein: the mutated mouse was unable to develop, and the embryo died rapidly. A similar defect in embryogenesis was demonstrated by knock-out of the genes encoding fibronectin and collagen, which participate in formation of the extracellular matrix.

However, knock-out does not always yield such clear-cut results. For example, Lesch–Nyhan disease is caused by mutation of a gene encoding an enzyme involved in synthesizing nucleotide components, and in humans this provokes a characteristic dementia. Obtaining mice mutated

Figure 6.10 Knock-out in the mouse. (A) The gene of interest M is inactivated in ES cells of an *agouti* line by targetted insertion: it is replaced by a mutant gene m (or a resistance gene such as neo, conferring resistance to neomycin), by homologous recombination between the sequences that flank it and those which flank the wild-type gene M. (B) The ES cell is introduced into a blastocyst, and the construct is implanted into a pseudo-pregnant mouse. Chimeric mice are obtained in the Fl, containing both wild-type and agouti cells. A chimeric mouse whose germline is *agouti* (and thus M/m) is then crossed to a wild-type mouse (M/M), and from the F2, heterozygous mice (M/m) are obtained. Crossing two of these heterozygous mice gives an F3 generation in which m/m double mutants are produced in one-quarter of cases (unless the homozygous mutant is lethal, in which case one obtains two-thirds M/m heterozygotes and one-third M/M wild-types)

in this protein showed no evidence for similar pathology (possibly because the gene has different functions in the two species). Knock-outs of the gene encoding tenascin-C, a protein implicated in cellular migration, or of the gene for vimentin (intermediate filament), have no effect, although their role is considered to be vital. Perhaps this role is compensated by redundancy of the mutated gene within the genome, or is not detectable under the growth conditions traditionally employed.

To date, about 1000 knock-outs have been obtained in the mouse. A private American company – Lexicon Genetics – has even been created, which offers a collection of several thousand directed mutations of the mouse.

6.3.4   Random mutagenesis

Random mutagenesis may be carried out using mutagens such as ethyl-methane-sulphonate (EMS), trimethyl psoralen (TMP/UV) or ethyl-nitrosourea (ENU). This approach is easily carried out in prokaryotes; for example, 1500 *B. subtilis* genes of unknown function have thus been inactivated.

In eukaryotes, this strategy was greatly developed during the 1980s through extensive research on mutations affecting embryonic development in *Drosophila*. This work was carried out by Nüsslein–Volhard and Wieschaus, and led to the identification of a large number of genes encoding proteins required for establishing the dorso-ventral and antero-posterior axes, segmentation, and the control of segment identity during development. These workers received the Nobel Prize for physiology and medicine in 1995.

In *C. elegans*, many phenotypes have also been obtained, and several hundred genes encoding proteins involved in diverse functions have been identified: these include muscular differentiation (e.g. myosins, paramyosins and actins), cellular development, neurolocomotory system, sensory functions (chemotaxis, thermotaxis, olfaction, mechanoreception), behaviour (feeding, defecation, copulation, egg-laying), neuronal plasticity and drug resistance.

Another approach is to use a transposon. In yeast one may use the Ty1, mini-Mu or Tn3/5 transposons. The advantage of a transposon is that the mutated gene may be rapidly identified, by exploiting the known sequence of the inserted transposon. In *C. elegans*, a gene may be mutated using a type Tc1 transposon, followed by imprecise excision of the element. Concommitant with the excision part of the coding region of the gene into which the Tc1 was inserted may be deleted, thus mutating this gene.

In *Arabidopsis*, a large number of mutant lines have been developed by insertion of a transposable element from maize (Ac/Ds or En/Spm). The mutated gene may be identified using the known sequence of the inserted element. Such collections may be used to recognize the mutant phenotype of interest; the insertion mutant is then identified, which allows isolation of the implicated gene and characterization of the function of the protein it encodes.

Another strategy has been developed in *A. thaliana*: it consists of using the Ti plasmid of *Agrobacterium tumefaciens*, which is capable of inserting the sequence flanked by the two DNA-T regions into the genome of *A. thaliana* (Figure 6.8). Since this genome consists mainly of coding sequences, a large fraction (19 per cent) of these transformants show a particular phenotype when homozygous. After random insertion, the plant is self-fertilized. A quarter of the subsequent generation represent non-mutated homozygotes, half are heterozygotes, and a quarter are double mutants (provided the mutation is viable; if it is lethal the ratios are 1/3–2/3). The mutated genes are subsequently identified by PCR.

The phenotypic modifications induced by mutating a gene, and thus the protein it encodes, may be obvious, or on the other hand difficult to detect, necessitating very precise culture conditions. For example, mutation of the gene encoding the AKT1 potassium transporter is undetectable except when potassium is absent from the culture. For some mutations no modification is detected, probably because of genome redundance.

In the mouse, more than 1000 mutations have been reported, and the mutated gene has been identified in 130 cases (45 per cent correspond to a human mutation that induces a genetic disease). However, in vertebrates, research on the genes of development has reached its apogee with the work of Nüsslein–Volhard and Driever, who have carried out a programme of random mutagenesis on the fish *Danio rerio* using ENU. More than one and a half million embryos have been observed, and 1858 mutations affecting development have been identified.

6.4 Two-dimensional Electrophoresis and Identification of Proteins

The study of proteins has undergone a spectacular evolution. Two-dimensional electrophoresis and mass spectrometry effectively allow a global approach to the identification of the proteins present in a given tissue. This type of study is completely detached from the analysis of transcription: tissues may be studied in which no transcript is present, for example blood serum or intercellular fluid.

6.4.1 Separation of proteins by two-dimensional electrophoresis

With a view to carrying out two-dimensional (2D) protein electrophoresis, cells or tissues are first of all disaggregated using non-ionic detergents, so as to solubilize the proteins. Proteolysis is avoided by using a low temperature (4°C) and agents such as EDTA or cocktails of protease inhibitors. Insoluble substrates are removed by centrifugation, and reducing agents such as dithiothreitol (DTT) are employed to eliminate disulphide bonds.

The first separation (on the X-axis) is carried out as a function of the isoelectric point of each protein (IEF, isoelectric focusing). The proteins are separated on a pH gradient (between pH 3 and pH 12) under the influence of an electric field. A protein ceases to migrate when it reaches the region where the pH neutralizes its electric charge (Figure 6.11A).

The second separation (on the Y-axis) is an SDS–PAGE-type electrophoresis using a polyacrylamide gel in the presence of an anionic detergent (usually SDS). SDS imposes an equivalent charge on most proteins, so the separation depends only on mass (Figure 6.11B). This separation is effective for proteins of mass between 10 and 200 kDa (some approaches allow the separation of larger or smaller sizes).

After separation, the proteins are detected by staining, for example using silver nitrate or Coomassie blue (Figure 6.12). Coomassie blue only allows the detection of proteins present at more than 100 ng, but silver staining is 10–100 times more sensitive. The images thus produced are captured by informatic acquisition systems, and it is then possible to establish databases containing many such images, and to compare them.

By this approach, several thousand proteins of a biological system may be analysed in a quantitative manner, which allows the establishment of 2D translation maps. However rare proteins, such as transcription factors or signal proteins, are often the most interesting, and may be swamped by more abundant proteins (often housekeeping proteins). The first are present in 10–1000 copies per cell, the second at more than 10 000 copies.

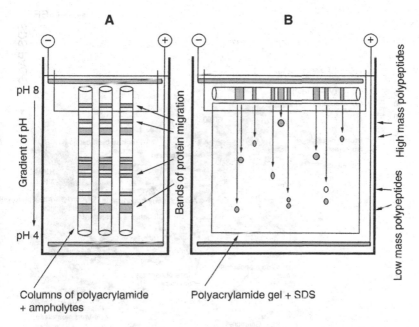

Figure 6.11 Migration of proteins in a 2D gel. (A) IEF migration: cell extracts are placed on a gel column whose pH is determined by ampholytes, and the proteins migrate in the electric field to the point where the pH cancels their charge. (B) SDS–PAGE migration: after IEF migration, the column is placed sideways onto an SDS gel, and migration takes place according to molecular weight. Calibration of molecular weight and pH is obtained using protein standards which have migrated under the same conditions

6.4.2 Protein identification by classic techniques

How, starting from 2D gels, can one identify a protein thus isolated with certainty? Thanks to progress in genomic sequencing and the amount of proteic data available, location of a protein after migration sometimes allows the identification of the protein through database searches. Yet such searches are rarely fruitful.

One experimental approach is the Western blot, which allows the identification of a given protein by means of an antibody which recognizes it specifically: in this case, the proteins of the 2D gel are transferred to a membrane, and exposed to the antibody specific for the protein of interest (a membrane may be used for this purpose several times). However, such an approach is not practical on a large scale, because only a small number

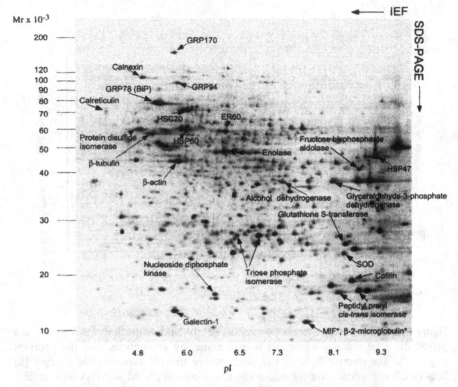

Figure 6.12 Example of 2D migration. Each spot represents one (or sometimes several) stained peptides. In this case they originate in hamster ovarian cells. A certain number of them which have been identified are named

of proteins can thus be identified by antibodies, and reduced or denatured proteins are not always recognized by specific antibodies.

Another approach is to establish the amino-acid composition of a protein. The protein is first isolated from the gel, hydrolysed, and then fluorescence chromatography is used to evaluate the relative abundance of each amino acid in the protein. In some cases these data suffice for its identification.

N-terminal protein sequencing of the isolated protein may also be performed using the method of Edman, and this approach can give the sequence of the first 10–15 amino acids, which often leads to the identification of the protein (Table 6.4). However, this technique is expensive, and numerous eukaryote proteins have blocked N-termini, so are not sequenceable. This approach is thus not applicable on a large scale.

Table 6.4   Examples of proteins identified through N-terminal sequencing

N-terminal sequence	Protein identified
SKIFEDNS	Cysteine synthetase A
MKVAVLGAAGGIGQAL	Malate dehydrogenase
FPTIPL	Growth hormone
MNIRPLLHDRVIVKRKEVE	Chaperonin (10 kDa)

6.4.3   Identification of proteins by mass spectrometry

Recently techniques have emerged which rely on a different principle: mass spectrometry (Figure 6.13). This consists of isolating a protein from a gel after 2D separation, digesting it with an endoprotease such as trypsin (which cuts the peptide bond on the carboxyterminal side of lysine or arginine) or chymotrypsin (which cuts on the C-terminal side of hydrophobic amino acids), so as to produce peptide fragments whose mass may be determined by MALDI-TOF MS (matrix assisted laser desorption/ionization–time of flight mass spectrometry). The digestion products are co-crystallized with a solid phase matrix, presented on a metallic surface to a laser, which desorbs them, and the mass of each peptide is given by its speed of flight towards the detector (Figure 6.14).

The original protein may thus be identified, by comparison with protein or genomic databases (Figure 6.15). With three to six digestion peptides it is possible to identify a protein in *E. coli*; four to six suffice for yeast. Large proteins may not be identifiable because of the large numbers of peptides obtained. This approach may also be compromised by the presence of several proteins in the gel fragment analysed, or by post-transcriptional modifications.

Another approach is ESI-MS/MS (electrospray ionization). As before, a protein is separated by 2D electrophoresis, isolated and digested by an endopeptidase. The products of this digestion are then sorted using four bar magnets forming a first quadripole, whose polarization is switched at high frequency, which determines the charge selected in the axis. Other charges diverge, and are eliminated. This corresponds to precise selection of a MALDI-TOF peak. After this selection, the sample is fragmented by collisions with a neutral gas, such as argon or xenon (CID = collision induced dissociation), a second quadripole separates on the basis of size, and a third allows analysis (Figure 6.16).

The Y extremity, which has a positive charge on the nitrogen terminus, is analysed. A first series of signals is thus detected. The B extremity

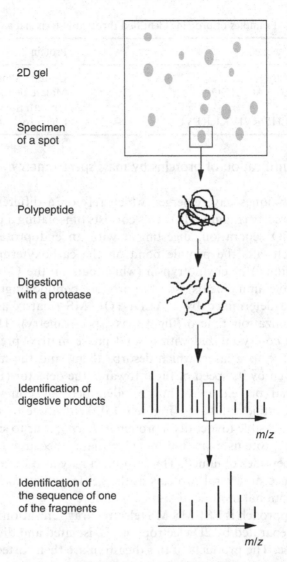

2D gel

**Specimen
of a spot**

Polypeptide

**Digestion
with a protease**

**Identification of
digestive products**

m/z

**Identification of
the sequence of one
of the fragments**

m/z

Figure 6.13 Analysis by MALDI-TOF or ESI-MS/MS. The polypeptides are first separated in a 2D gel. One of them is isolated, and digested by a protease. In MALDI-TOF the peaks obtained represent the resultant peptides (m/z = mass/charge ratio). Assembly of these data may allow identification of the protein. In ESI-MS/MS the protein sequence may be deduced from the digestion products

(carboxy-terminal) is then made positively charged, and a second series of signals is detected. Comparison between the two peaks gives the mass of the amino acid which has been lost, thus allowing its identification. Analysis of a digestion product thus permits the identification of the

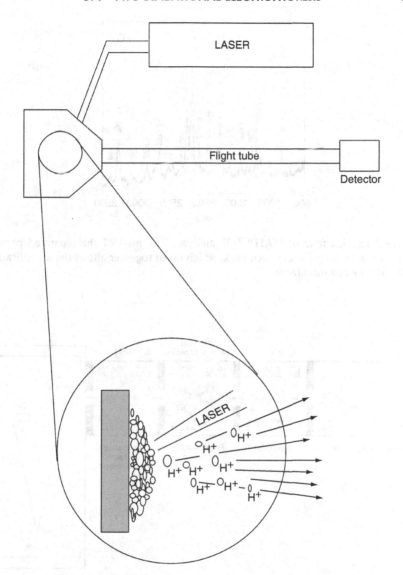

Figure 6.14 MALDI-TOF analysis. The products of digestion with endoprotease are exposed to a laser, charged by a proton H+, and the time of flight to the detector allows identification of the mass of each peptide

complete protein sequence of the peptide (Figure 6.17). Comparison with the databases then allows identification of the protein analysed (Table 6.5).

This technique lends itself to implementation at a large scale, permitting the automation of the production of protein sequences, and their identification. It is also extremely sensitive, since several femtomoles or even attomoles of protein suffice.

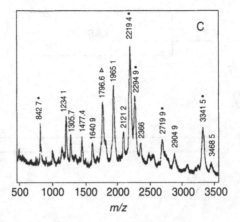

Figure 6.15 Example of MALDI-TOF analysis. The mass of the separated protein fragments is indicated above each peak, which taken together allows the identification of a 60 kDa elongation factor

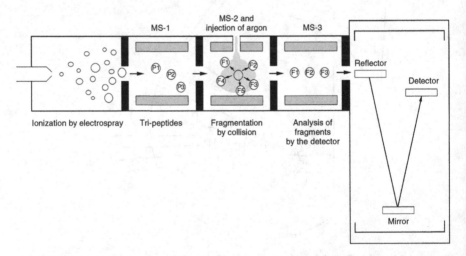

Figure 6.16 ESI-MS/MS analysis. Ionization occurs in an electric field which will positively charge the peptides obtained by digestion (charge of one proton). The first quadripole (MS-I) allows the selection of one of the peptides (P2) as a function of the ratio between its mass and charge (m/z). The selected peptide then passes through an inert gas (in this case argon) thanks to a second quadripole (MS-2), and the collisions fragment the peptide (the fragmentation mainly affects the amide bond). Thus a series of peptides is obtained whose charges differ because of the loss of one or several amino acids (FI, F2, F3 etc.). The third quadripole (MS-3) allows one to obtain the m/z ratio for each of these, which allows the final deduction of the protein sequence

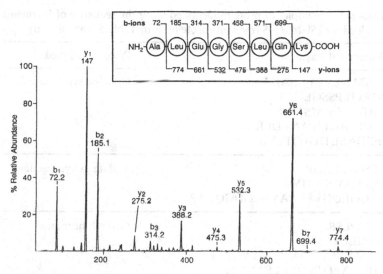

Figure 6.17 Example of ESI analysis. The peaks b represent the fragmentation results obtained when the amino-terminal is positively charged, and the peaks y those obtained when the carboxy-terminal is labelled. The charge differences allow the determination of which amino acid has disappeared. For example, $b3 - b2 = 314 - 185 = y6 - y5 = 661 - 532 = 129$, which corresponds to glutamic acid

6.4.4 Examples of applications

Amongst the most-studied areas in humans, we may cite the identification of proteins specifically expressed during various syndromes, such as hypertension, ageing, myocarditis and certain cancers. Studies on cerebro-spinal cells of Alzheimer's disease patients have led to the identification of proteins whose expression changes during the course of this syndrome. This approach has also allowed the study of the effects of medicines, either pre-existing or novel; for example the effect of herceptin, developed to fight breast cancer, has been analysed by this method. Spectrometry also allows the identification of drugs capable of interacting with proteins implicated in a given disease. This approach thus permits the testing of a large number of drugs, and the rapid establishment of how they interact with the protein of interest. Thus thrombin, implicated in the disease thromboembolism, has been examined by ESI to study the effect of an acylating drug.

Studies on pathological species are principally concerned with *Salmonella*, *M. tuberculosis*, *V. cholerae*, *M. genitalium* or *H. influenzae*. Additionally, comparison between related species, such as *S. typhimurium*

Table 6.5 Examples of proteins identified using the sequence of fragments obtained by ESI–MS/MS (the same proteins as in Table 6.4 are shown)

Sequence of fragments	Protein identified
ALGANLVLTEGAK NIVVILPSSGER VIGITNEEAISTAR IQGIGAGFIPANLDLK IFEDNSLTIGHTPLVR	Cysteine synthetase A
FFSQPLLLGK SKLFNVNAGIVK ALQGEQGVVECAYVEGDGQYAR	Malate dehydrogenase
FPTIPLSR NYGLLYCFR SVFANSLVYGASNSDVYDLLK LHQLAFDTYQEFEEAYIPK	Growth hormone
EMLPVLEAVAK GQNEDQNVGIK AAVEEGVVAGGGVALIR AIAQVGTISANSDETVGK ANDAAGDGTTTATVLAQAIITEGLK EGVITVEDGTGLQDELDVVEGMQFDR	Chaperonin (10 kDa)

and *E. coli*, should permit the identification of the virulence factors specific to the first, pathogenic, species. Other non-pathogenic species are studied, for example *Synechocystis* sp., *Escherichia coli*, *Rhizobium leguminosarum*, *Dictyostelium discoideum*, *S. cerevisiae*, *B. subtilis*, mouse and rat. Some of these have been studied under variable environmental, physiological or metabolic conditions.

Protein modifications may also be studied by mass spectrometry. About a quarter of proteins undergo such modifications, which change their molecular weight, and spectrometry allows the identification of some of these, by comparison of the observed and expected masses for a given protein. For example, phosphorylation is reflected by a gain in mass of 80 Da, and acetylation by a gain of 42 Da.

6.4.5 Protein interactions

At the biochemical level, proteins practically always interact with other proteins. The TAP (tandem-affinity purification) approach consists of inserting 3′ to a gene encoding a yeast protein ('bait' protein), a sequence encoding a protein region which is to be used as a marker, and introducing this construct into the yeast genome. Each clone thus expresses this hybrid protein, and the collection of proteins interacting with it ('prey' proteins) is purified, separated on SDS–PAGE and analysed by mass spectrometry after trypsin digestion. Since the complete genomic sequence of yeast is known, the protein data obtained by spectrometry allow the identification of which protein has thus been isolated in this way, and thus which proteins interact with the one whose gene was initially introduced into yeast (Figure 6.18).

This approach may be carried out on a large scale, at the whole proteome level, especially in yeast. Several hundred proteins have thus been identified, allowing the establishment of about 11 000 protein–protein interactions (Figure 6.19). For a quarter of these, no function had been previously proposed. Groups of proteins interacting with one another have thus been identified. On average these groups comprise 12 proteins, and 80 per cent of the interactions thus demonstrated were previously unknown. Many of these groups share common proteins.

Comparison of the information obtained in yeast with that known for humans shows that comparable interactions are found for proteins that are homologous between the two species. Thus the yeast data is to some extent transposable to humans.

6.5 Identification of Protein Interactions by Two-hybrid system

6.5.1 Strategy

Protein interactions are the basis of a large number of biochemical reactions, such as the formation of protein complexes, molecular machines and cellular macrostructures (cytoskeleton, mitotic spindle, chromosome structures and centrosomes), of enzyme–ligand interactions, and of transcriptional regulation (promoter–regulator protein interactions). The study of such interactions has been facilitated by the development of the two-hybrid system, which allows the identification of the proteins which interact with one another (or which interact with DNA or another protein in the case of the triple-hybrid). It even permits the precise identification of the domains involved.

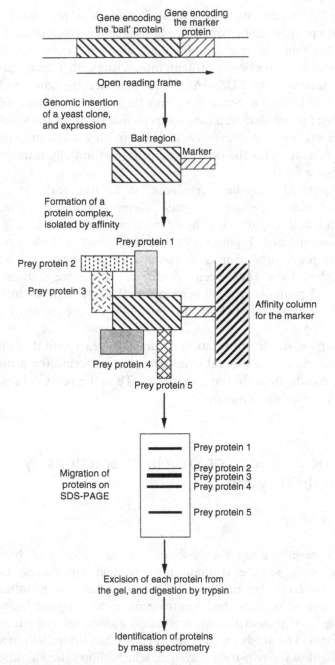

Figure 6.18 Studying protein interactions by spectrometry. A chimeric gene comprising the sequence encoding the 'bait' protein and that of the marker-protein is constructed, and inserted into the yeast genome. This clone expresses the chimeric

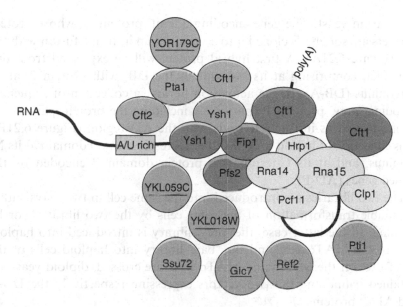

Figure 6.19 Proteins involved in messenger polyadenylation in yeast. The 3' extremity of a message is represented, along wth 21 proteins involved in polyadenylation. Proteins newly identified by spectrometry are underlined

The strategy was described in 1989, the protein used being the transcriptional activator GAL4. This protein comprises two domains: a DNA binding domain (DB, DNA binding domain) and an activation domain (AD, activation domain). If this two-domain protein is present in the nucleus, the DB specifically recognizes a region of DNA located in a promoter (UAS, upstream activation sequence), and the AD activates transcription of the gene controlled by that promoter (Figure 6.20A). These two domains are active if they are on the same protein, but inactive if separated into two independent peptides (Figure 6.20B and C). The remarkable fact is that the independent AD and DB peptides may become active, and activate transcription, if a system rejoins them within the nucleus, even if no covalent interaction is involved (Figure 6.20D).

This has led to the development of a strategy allowing the identification of protein interactions, by reconstitution of a transcriptional

protein (bait), with which other (prey) proteins interact, and this complex is selectively recovered using the marker-protein. These proteins are separated by electrophoresis, and each of the bands is analysed by spectrometry. For each of them one (or several) peptide sequences is obtained, and comparing these sequences with the complete yeast genome allows the identification of each 'prey' protein activator in yeast. The gene encoding a 'bait' protein A, whose protein partners are sought, is cloned into a vector as an in-frame fusion with the DB (Figure 6.21A). A first hybrid protein will be expressed from this construct, comprising at its N-terminus the DB, with protein A at its C-terminus (DB-A). In the activation library, a collection of sequences encoding 'prey' proteins P (putatively including the protein partners of A) are cloned as in-frame hybrids with the AD region (Figure 6.21B). This collection of hybrid proteins thus expresses the AD domain at its N-terminus, and at its C-terminus a protein domain P encoded by the cloned genes (AD-P).

The two libraries are introduced into the same cell in two ways: either by double transformation of the same cells by the two libraries, or by crossing. In the latter case, the 'prey' library is introduced into haploid cells of the MATα sex, and the 'bait' library into haploid cells of the MATa sex. If the MATa and MATα lines are crossed, diploid yeast are obtained containing the two vectors expressing respectively the DB-A and AD-P proteins.

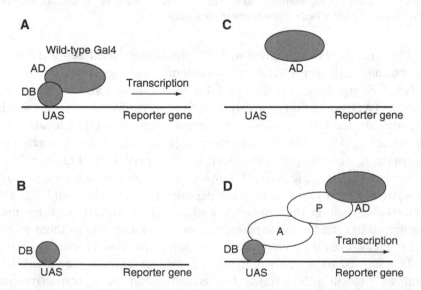

Figure 6.20 Two-hybrid system. (A) The Gal4 protein of yeast comprises two domains DB and AD, which activate transcription, because of the recognition of the UAS by DB, and transcriptional activation by AD. The reporter gene encodes a lacZ-HIS3 fusion. (B) If only the DB domain is present, transcription is not activated. (C) If only the AD domain is present, transcription is also not activated. (D) If the domains are separated, but each is joined to proteins A and P which recognize one another, the capacity for activation is restored, and the reporter gene is transcribed

The nucleotide sequences cloned in each expression vector must be directly translatable: thus they are cDNAs for studies on eukaryotes, or genomic sequences for prokaryotes. If interactions between yeast proteins are being studied, genomic sequences may also be used, because 68 per cent of the genome is coding. The diversity of clones in the library containing the AD domain should span the whole genome: for example, it should be of the order of several thousand clones for yeast.

Cells showing interaction between the A and P proteins are selected using reporter genes, whose transcription – and thus translation – is dependent on a promoter containing the UAS recognized by the DB domain. If the protein A interacts with the protein P, the AD and DB domains contained within the DB-A and AD-P hybrid proteins will be brought close to one another, and transcription of the reporter genes will be activated. The reporter gene HIS3, encoding an enzyme for histidine synthesis, allows nutritional selection for positive clones using a culture medium which does not contain histidine. The reporter gene lacZ encodes ß-galactosidase, and clones expressing positive two-hybrid thus exhibit a blue colour when cultivated in the presence of X-Gal. Vectors are extracted from positive clones thus identified, and sequencing the region inserted then allows identification of the P domain which is capable of interacting with protein A.

Failure to detect genuine protein interactions using the two-hybrid system (false negatives) may be caused by: (i) a toxic effect of a hybrid protein; (ii) the absence in yeast of indispensible post-transcriptional modifications (such as tyrosine phosphorylation); (iii) the production of very hydrophobic proteins (e.g. transmembrane domains); (iv) instability or incorrect conformation of the proteins when expressed in yeast; and (v) inability of the hybrid proteins to traverse the nuclear membrane so as to activate transcription in the nucleus. False positives may also be obtained, reflecting interactions within yeast which do not correspond to any real interaction in the organism studied. Finally, these tests rely on transcription: transcription itself may therefore not be analysed.

6.5.2 Protein–protein interactions

The two-hybrid system may be applied to the study of possible interactions in a limited collection of proteins amongst which interactions are supposed to occur, or to interactions between one protein and a large number of other domains obtained from the transcribed sequences in a developmental stage or given tissue.

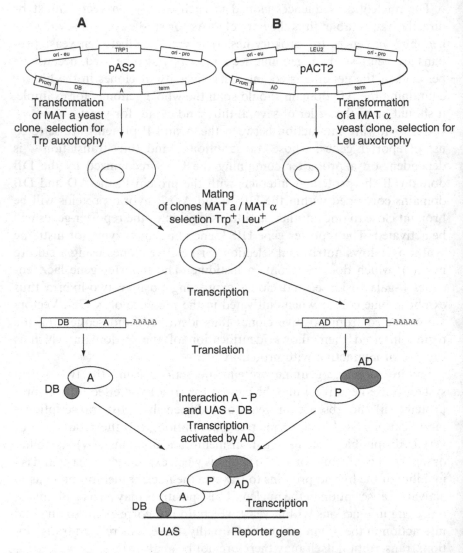

Figure 6.21 Technique of the two-hybrid system. (A) The DNA region encoding the DB domain of Gal4 (the first 147 amino acids of the N-terminal region) is linked with that encoding protein A in the pAS2 vector, and MATa yeast is transformed by this construct. The vector propagates in yeast by means of an eukaryotic origin of replication (ori-eu), and the recombinants are selected by culture in a tryptophan-free medium (the TRP1 gene encodes a enzyme which allows yeast to synthesise this amino acid for itself). The promoter (Prom) induces nuclear transcription of the DB-A gene-hybrid up to the termination site (term), and the gene is translated in the cytoplasm. (B) The region encoding the AD domain of GAL4 (amino acids 768–881 in the C-terminal region) is associated with that encoding the protein P in the pAC2

For example, analysis of interactions between actin and its ligands has been carried out using a two-hybrid library containing the gene which encodes this protein, mutated in a random fashion. Three types of consequences were observed: (i) clones for which all the ligands interacted with the mutant actin (Figure 6.22A); (ii) clones for which no reaction was detectable (Figure 6.22B); and (iii) clones where the mutant actin interacted with some ligands but not with others. The first case suggests that the three-dimensional structure of actin has not been modified. In the second case, it is probable that the structure is unstable or is completely different from that of wild-type actin. The last case is the most interesting, giving considerable information on the actin–ligand interaction. When the results are compared with the three-dimensional structure of actin, one observes that the mutants which modify the interactions fall into different and well-separated regions.

In yeast, functions may be proposed for proteins which interact with proteins of known function; for example, two proteins have been associated with arginine metabolism, because of their interaction with ornithine amino-transferase (an enzyme implicated in arginine biosynthesis). Previously unknown interactions between proteins implicated in the same function have also been detected. Thus, evidence has been found for novel interactions amongst regulators of the cell cycle, or proteins mediating vesicle transport (Figure 6.23A). Interactions between proteins involved with different functions may also be detected, thus establishing previously unsuspected correlations between diverse pathways; for example, interactions between proteins involved in crossing over, and double-strand breakage of DNA, have been detected.

6.5.3 RNA–protein interactions

The efficiency of this system also allows the identification of interactions between various proteins and RNA: if a 'bait' protein specifically recognizes RNA, one may identify 'prey' proteins that recognize the same RNA. In this type of research, A is associated with AD, the RNA is transcribed

---

vector, and the construct is used to transform MATα yeast. The recombinants are selected by culture in a leucine-free medium (the LEU1 gene encodes an enzyme allowing synthesis of this amino acid). The AD-P gene hybrid is transcribed and translated. The MATα and MATa clones are conjugated, and the positive clones are identified through expression of the reporter gene; the vectors are then extracted, amplified in a bacterial system (by means of the bacterial origin of replication ori–pro), and sequenced. The sequence obtained allows the deduction of that of the protein P

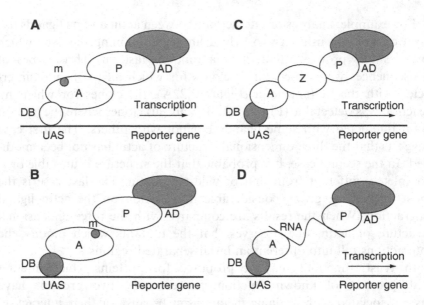

Figure 6.22 Studying interactions by means of two-hybrids. (A) Where A and P are known, the site of interaction between these proteins may be studied by producing mutations (m) in protein A: if this mutation does not affect A-P recognition, the reporter gene is transcribed. (B) If the mutation modifies A-P recognition, the reporter gene is not transcribed. Studying interactions by means of triple-hybrids. (C) Where A and P are known, but do not interact directly, a protein Z which interacts simultaneously with both A and P may be identified by means of a triple-hybrid, which brings together DB and AD and activates reporter gene transcription. (D) Where A interacts with an RNA, other proteins P which recognize this RNA may also be identified using the triple-hybrid system

from a separate vector, and P is associated with AD (Figure 6.22D). If the RNA is recognized simultaneously by A-DB and P-AD, DB and AD are brought into proximity, and activate reporter gene transcription. Sequencing the vector encoding P-AD then allows identification of the protein P. This strategy, involving three molecules, is called triple-hybrid.

This approach has been applied to identify novel proteins involved in RNA recognition during splicing, translation, or retrovirus infections such as HIV.

6.5.4 Global approaches

The advantages of the two-hybrid approach are that no protein purification is required: identification of the interacting proteins is finally obtained by nucleic acid sequencing. Transient interactions may also be identified, whereas traditional techniques only allow the identification of stable interactions. Large-scale explorations have been undertaken, covering all the proteins encoded by a genome (Table 6.6), which has allowed the establishment of a network of complex protein interactions (Figure 6.23B).

In prokaryotes, the small size of the genome means this kind of research is relatively easy. Thus for *H. pylori* it has been possible to study all the proteins interacting with 261 proteins of interest, and 1280 interactions have been identified, for example those involved with chemotaxis, DNA or RNA synthesis, or the urease metabolic complex (important for pathogenicity).

The global approach is also practicable in yeast. Some of the protein interactions identified are for example implicated in messenger splicing, in the formation of the synaptonemal complex, in spindle pole formation, ribosomes, sporulation or vesicular transport. Many of these proteins had not previously been functionally characterized. Overall, an interaction has been identified for 8 per cent of proteins, and half these interactions are novel.

For other eukaryotes this approach is more difficult, because not all the genes are identified with certainty, they are often interrupted by introns, and the cDNA libraries cannot be considered to represent the entire coding fraction. It has nevertheless been carried out for the nematode: 29 proteins implicated in vulval development have been studied, which has led to the identification of 124 proteins involved in 148 interactions, of which only 15 were previously known.

Table 6.6 Examples of protein interaction studies carried out for various completely sequenced species

Organism	Total number of genes	AD-P library	DB-A library	Interactions identified
Smallpox	266	All genes	All genes	37
Helicobacter pylori	1550	All genes	261 genes	1280
Yeast	6200	5345 genes	5345 genes	692
C. elegans	19 100	All genes	27 genes	148
D. melanogaster	13 600	5 genes	9 genes	19

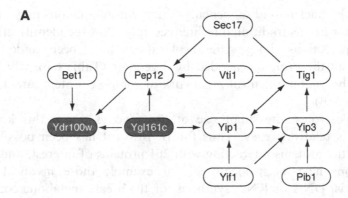

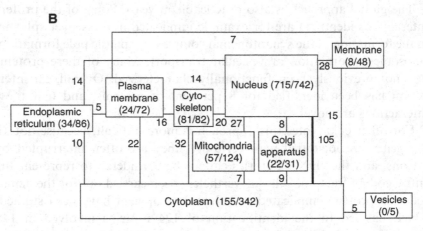

Figure 6.23 Protein interactions detected in yeast. (A) Examples of interactions detected amongst proteins implicated in vesicular transport. (B) Total protein interactions detected in different cellular compartments. In each compartment is indicated the number of proteins for which evidence of interaction has been obtained, divided by the number of proteins tested, and the figures indicated between two different compartments give the number of interactions detected between them

6.6   Chip Technology

6.6.1   Construction of protein chips

We saw in Section 5.5.2 that DNA chips have been employed with great success, both in the field of re-sequencing and in research on tissue-specific expression profiles. This has inspired the development of similar approaches in the protein domain, in particular using chips upon which a collection of proteins is fixed.

These pose technical problems because (i) proteins consist of 20 distinct amino acids, whilst there are only four bases in DNA, (ii) depending on their amino-acid composition, proteins may be hydrophilic, hydrophobic, acidic or basic (whilst DNA is always hydrophilic and negatively charged), (iii) proteins are often post-translationally modified (by glycosylation, phosphorylation, etc.).

Nonetheless protein chips have been made: the proteins are deposited onto a support and subsequently fixed to it. One may thus arrange 1600 distinct proteins per cm^2. These arrays are ordered, so that one knows which protein is represented by a given spot. The chips are now incubated with other ligands (fluorescently labelled), and the results of hybridization are analysed by confocal microscopy (it is also possible to employ radioactively labelled ligands). The proteins recognized may be identified, using the signal localization data obtained. The intensity of the signal obtained is proportional to the level of ligand–protein interaction.

6.6.2 Strategies employed

Protein chips have been used for various applications. One consists of fixing onto the chip a collection of proteins amongst which some are suspected of being ligands for an enzyme (Figure 6.24A). For example, this was carried out during research on kinase ligands in *S. cerevisiae*. The chip was incubated in the presence of the enzyme and ATP, whose phosphorous was radioactively labelled. After washing, a collection of labelled proteins was identified, showing that these were possible ligands for this enzyme. Some of these had not previously been identified as such.

Another case is the production of chips carrying a large number of small proteins, randomly synthesized *in vitro* (Figure 6.24B). These chips may for example be hybridized to a protein of the receptor type, for which novel ligands are sought. This may facilitate the eventual identification of competitors for the receptor.

Chips have also been produced to which a large collection of antibodies have been fixed. For each antibody, the specifically recognized protein was known. All the proteins from a given tissue were extracted, labelled, and incubated with this chip. A large repertoire of proteins may thus be analysed, and their abundance within the tissue may be estimated as a function of the signal intensity obtained with each antibody.

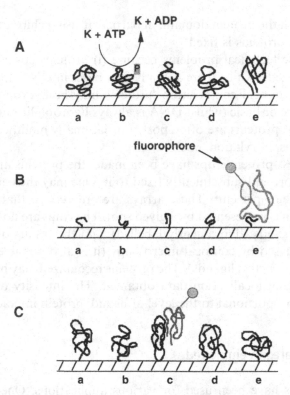

Figure 6.24 Protein chips. (A) Detection of a kinase activity using proteins fixed onto a chip. In position b, the fixed protein is a kinase substrate, which results in linkage to a phosphate, which may be detected if the phosphate is radioactive. (B) Detection of a new ligand for a receptor. In position e, one of the mini-proteins fixed onto the chip is recognized by the receptor. Labelling of the latter by a fluorophore allows the identification of the mini-protein recognized. (C) Large-scale detection of protein interactions. Proteins are fixed onto a chip, and one of these, in position c, is recognized by the protein-probe being tested. The probe protein is labelled with a fluorophore, which allows the identification of the protein recognized

Finally, chips carrying virtually all the proteins of yeast (80 per cent) have been produced (Figure 6.24C). This 'complete proteome' approach is employed with the aim of identifying all possible interactions between a given protein and the collection of proteins which exist in this species. In this way, for example using calmodulin as a 'protein-probe', it has been possible to identify proteins already known to interact with this molecule, and also new proteins, previously unsuspected of such interactions.

6.7   Analysis of Three-dimensional Structure

A protein is an assembly of amino acids whose sequence is determined by that of the gene which encodes it. The analysis of three-dimensional structure of a protein (meaning the shape adopted by this assembly in space) is a major asset for understanding its function; for example, how an enzyme is capable of catalysing a specific chemical reaction, how a receptor recognizes a ligand and transmits a message, and how one protein in a structure interacts with other components (intra- or inter-cellular). Analysing three-dimensional structure is costly, but it is currently impossible to predict 3D structure from a protein sequence. To achieve this, two approaches are employed: crystallography and spectroscopy, which rely on the analysis of radiation after it has traversed a molecule.

6.7.1   Crystallography

To obtain structural information by crystallography, the proteins must first be crystallized. The crystals are then illuminated with X-rays, whose wavelength is about 1 Ångström, the same order of magnitude as the inter-atomic distances within the protein being studied (neutrons and electrons can be used in some cases).

As the wave passes through the crystals, a fraction of it is diffracted in different directions. Its wavelength does not change, but its amplitude and phase are modified. All the diffractions obtained are the same for each crystal. This is known as crystal diffraction. If a lens system were available for X-rays (which is not the case), it would be possible directly to refocus the diffracted waves so as to obtain an enlarged image of the crystalline structure, as is done in the optical microscope. In X-ray crystallography, the modifications of the waves diffracted by the crystal are detected as a collection of spots, and analysis by Fourier transformation then permits the reconstruction of the structure of the illuminated crystal.

The first protein to be thus analysed was myoglobin, by Kendrew and Perutz in 1965. Today, more than 14 500 proteins have been analysed by crystallography, and in the majority of cases (86 per cent) X-rays were employed.

6.7.2 Nuclear magnetic resonance

In nuclear magnetic resonance (NMR), one studies the exchange of energy between a wave and the atom it encounters. For this approach, it is not necessary to employ a crystal: a solution suffices (however, only small proteins may be studied, less than 30 000 Da).

NMR is carried out as follows: the sample, maintained in a constant magnetic field, is illuminated by waves of diverse frequency, and the resonances are then observed. Owing to interference between a nucleus and its environmental atoms, the frequency of the resonance is slightly displaced. This phenomenon allows structural studies, and analysis of the resonances permits the reconstruction of the three-dimensional structure of the protein. Generally one thus obtains a family of globally similar structures.

6.7.3 Protein structure

Two essential domains of protein secondary structure have been identified, the α helix and the β sheet. In the α helix, the oxygen in the peptidebond (C=O) is linked by a hydrogen bond to the hydrogen (N–H) of the fourth peptide bond. The two groups C=O and N–H of the same peptide bond are thus each implicated in a hydrogen bond, which is correlated with the formation of a peptide helix, each turn of which comprises 3.6 amino acids, the side groups being located on the exterior of this helix.

In the β sheet, hydrogen bonds are established between different regions of the same protein, or different proteins. According to the orientation of these regions, the sheet may be parallel (same N-terminal/C-terminal orientation in both regions) or antiparallel (opposite orientations). Here again, hydrogen bonds are established between the C=O and N–H groups of the peptide bonds in the two regions involved. The structures of the α helix and β sheet were predicted by Pauling in 1951, which won him the Nobel Prize for Chemistry in 1954.

Several types of assembly may be recognized in the three-dimensional spatial structures of proteins, at the heart of which amino acids arrange themselves in a restricted manner:

- in fibrous proteins, such as collagen, the structures are very elongated;

• in soluble globular proteins, the hydrophobic side chains are generally surrounded by a region carrying hydrophilic side chains;

• in membrane proteins, a hydrophobic segment contacts the lipids, and hydrophilic domains are situated on both sides of it;

• domains of α type are composed of α helixes, and are frequently encountered, for example in myoglobins;

• domains of β type are composed of β sheets, often antiparallel; these domains are found in enzymes, transport proteins, and immunity proteins;

• domains of the α-β type alternate these two types of motif; these domains are the commonest, being observed equally frequently amongst enzymes, transport proteins or proteins which recognise nucleic acids.

6.7.4   Example: histocompatibility antigens

The adaptive immune system is found in all vertebrates (Figure 6.25A). It is based on antibodies produced by B lymphocytes, the T-cell receptor (TCR), and histocompatibility molecules. T-cell receptors are involved in the recognition of antigenic determinants presented by histocompatibility molecules, which are very polymorphic membrane-bound glycoproteins (the genes encoding these proteins are localized in the same region of chromosome 6, and termed the major histocompatibility complex, or MHC). Histocompatibility proteins exist in two classes: type I proteins and type II proteins. Type I proteins are expressed on the surface of the majority of the cells in an organism, and they present antigens of cellular origin, which are peptide fragments of about ten amino acids produced by degradation of intracellular proteins.

Type I histocompatibility proteins are heterodimers composed of an α heavy chain, comprising three extracellular domains α1, α2 and α3, associated in a non-covalent manner with a light chain, β2 microglobulin. The heavy chain inserts into the membrane by a transmembrane domain and has a cytoplasmic extension. Crystallographic analysis of this protein was achieved in 1987 by Bjorkman (Figure 6.25B). The figure shows that the amino-terminal domains $\alpha 1$ and $\alpha 2$ form a 'pocket', whose base is a β sheet, and whose walls are made of two α helixes (the $\alpha 3$ and $\beta 2$ domains have a structure like that of the constant domain of immunoglobulins). In this 'pocket' one observes a region of high electron density which is in fact the site of fixation of cellular antigens, indicating that this type I molecule was crystallized 'in flagrante delicto' of presentation.

Crystallographic study of a type I proteins thus elucidated the function of the presentation system in cytotoxic T lymphocytes (which express the transmembrane protein CD8, but not CD4). This system is particularly important in virus infection and tumourigenesis, because presentation of virus or tumour antigens leads to lysis of the presenting cell by the cytotoxic T lymphocytes which recognize it.

Type II proteins are made of two transmembrane chains α and β, and are normally only expressed in thymic epithelium, Langherans cells, macrophages and B lymphocytes. They present antigens of extracellular origin. B lymphocytes, whose antibody is specific for an antigen, then interact with Th lymphocytes on which they depend (Th, helper T lymphocyte, $CD4^+CD8^+$). These cells then produce cytokines which stimulate the activity of the B lymphocytes. Thus, although the frequency of a cell specific for a given antigen is low, its increase in number consequent upon activation allows it to play a central role. Crystallography of type II proteins has demonstrated that the presentation site is very similar to that of type I proteins, which explains the important role of Th lymphocytes in the surveillance of antigens of endogenous origin.

6.8   Conclusion

Transgenesis has allowed the elucidation of the function of a substantial number of proteins, through the expression preference of the genes which encode them, and analysis of gene regulatory systems by promoters is also achievable. Another application is the production of transgenic 'bioreactor' species, in which are obtained proteins of therapeutic or economic interest. Other applications are to genetic diseases; for example, in 1997 a pig transgenic model of human retinitis pigmentosum was created.

Two-dimensional electrophoresis allied to mass spectrometry today permits the rapid identification of a protein or peptide: 10 years ago, two or three proteins a year could be identified. Now, several hundred

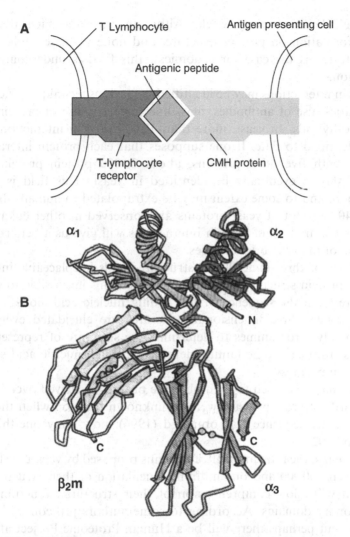

Figure 6.25 (A) Recognition of an antigenic peptide. The peptide is presented by an MHC protein on the presenting cell. It is recognized by a specific receptor expressed by a T lymphocyte. (B) Crystallographic analysis of an extracellular region of a class I protein. The $\alpha1$ and $\alpha2$ regions form a β sheet and two α helices, and the peptide is presented in this site. The $\alpha3$ region is associated with $\beta2$ microglobulin, and extended by a transmembrane and cytoplasmic region (not shown). The three disulphide bonds are shown ($\alpha2, \alpha3$ and $\beta2$ domains)

may be identified every week. Although this approach does not give information on protein structure, and none of these studies is yet complete (as is the case for genomics), this field is undergoing rapid expansion.

Protein interactions may be identified by various techniques, including informatics, use of antibodies, mass spectrometry, use of protein chips and two-hybrids. In yeast, more than 12 000 protein interactions have been identified to date. If one supposes that each protein interacts on average with five other proteins, about 30 000 protein–protein interactions should eventually be identified in yeast. This field is in full expansion, and to some extent may be extrapolated to humans, because about 40 per cent of yeast proteins are conserved in other eukaryotes. Perhaps the analysis of protein interactions will give us a better understanding of multifactorial diseases.

Analysis of three-dimensional structure yields irreplaceable information on protein structure and function (it remains impossible to predict structure from the sequence of the coding nucleic acid alone). In this field, six new three-dimensional structures are elucidated every day. Additionally, programmes to determine the structure of representative proteins from each gene family identified through nucleic acid sequencing are in progress.

Yeast may be chosen to exemplify the progress achieved over 6 years. Two-thirds of yeast proteins were of unknown function when the complete genome sequence was obtained (1996); now, only one-third are (1800 in 2002).

In humans, the total number of proteins proposed by various estimates is between 100 000 and one million. Elucidation of their structure and function will allow comprehension of their structural, functional and evolutionary domains. According to some authors proteomics is in its infancy, but perhaps there will be a Human Proteome Project after the Human Genome Project?

7 Identification of Genes Responsible for Disease

The setting-up of the Human Genome Programme has greatly improved the facilities for research on the genes responsible for human genetic disease. Such genes are numerous: McKusick has catalogued about 6000 monogenic disorders, the causative gene being known in only 18 per cent of cases. It is estimated that every human being carries five serious genetic anomalies. Happily these are generally not expressed, being masked by a non-mutant allele, but still one child in 100 is born with a genetic disease. Identifying the genes responsible for these diseases is a major challenge for medical research.

Isolating a gene responsible for a genetic disease is a complex problem: it relies on identifying mutations, perhaps alteration of a single base pair, within a genome that contains three million of them.

7.1 Genetic Diseases

7.1.1 Monogenic and multifactorial diseases

Monogenic genetic diseases are caused by mutations that affect a single gene. There are three types (Figure 7.1):

- Autosomal dominant diseases are caused by a mutation in a gene carried on an autosome, where the presence of a single mutant allele in an individual suffices to induce appearance of a disease (e.g. Huntington's chorea). This may be due to a gain of function in this

Genome, Transcriptome and Proteome Analysis   by Alain Bernot
© 2004 John Wiley & Sons, Ltd   ISBN 0 470 84954 1 (HB)   ISBN 0 470 84955 X (pbk)

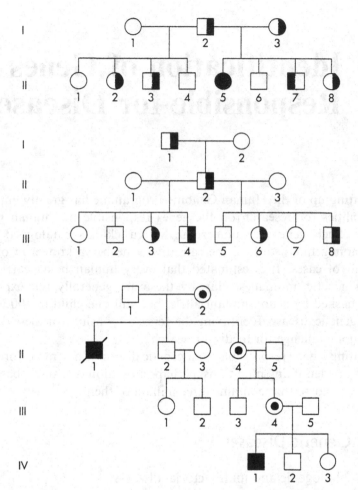

Figure 7.1   Modes of transmission of monogenic disorders. (A) Autosomal recessive genetic disease: starting from a pair of healthy carriers (individuals I.2 and I.3), a quarter of the children are affected (II.5), half are healthy carriers (II.7 and II.8); a couple comprising a healthy carrier (I.1) and a non-carrier (I.2) have no affected children, but half are healthy carriers (II.2 and II.3). (B) Autosomal dominant disease: amongst the descendants of an affected subject (I.1 or II.2), half the children are affected (III.3, III.4, III.6 and III.8). In autosomal diseases, males and females are equally affected. (C) X-linked disease: the daughters of a carrier female are not affected, but 50 per cent are carriers (II.4 or III.4). Fifty per cent of their sons are affected (II.1 or IV.1)

gene, or a dosage effect (expression of a single healthy allele being insufficient to ensure normal protein function).

- Autosomal recessive diseases arc also caused by a mutation in a gene carried by an autosome, but in this case both alleles must bc mutated for the disease to appear (e.g. cystic fibrosis); frequently therefore the defect is loss of gene function (individuals possessing only one mutant allele, and thus who do not manifest the disease, are termed healthy carriers).

- X-linked diseases manifest according to a particular pattern: in males, every mutation of a gene carried by this chromosome causes the appearance of the disease, because of the absence of a second X chromosome capable of compensating for the defect in the mutated gene (e.g. Duchenne muscular dystrophy). Females have two X chromosomes (one of which is inactivated in all cells by XIST, cf. section 1.1.5), so the disease appears in heterozygous females if the mutation is dominant, but only in homozygotes if it is recessive.

A certain number of diseases are associated with mutations in the mitochondrial genome. A mitochondrial disease is characterized by purely maternal inheritance, because of the particular mode of mitochondrial transmission. Effectively, only the oocyte transmits mitochondria to the zygote, from which all the mitochondria of the organism derive.

Additionally, there are a considerable number of multifactorial genetic diseases. These are caused by the interaction of multiple genes, and are influenced by environmental factors. The incidence of these diseases is very high: type I diabetes (IDDM, insulin-dependent diabetes mellitus) affects one child in 400; type II diabetes (NIDDM, non-insulin dependent diabetes mellitus) affects one person in 20; hypertension affects 15 per cent of the population.

7.1.2 Mutations

The variety of mutations capable of affecting a gene is vast. They manifest themselves by alteration of a translated protein, thus compromising its biological function, or by modification of its level of expression (Figure 7.2).

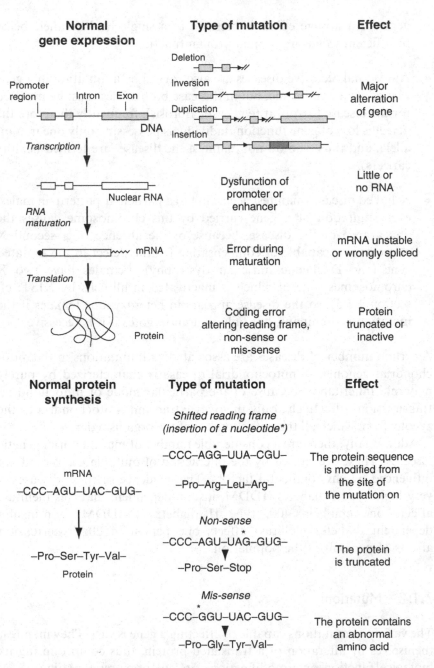

Figure 7.2 Types of mutation capable of affecting a gene, compared to normal gene expression

Chromosomal rearrangements (deletion, inversion, duplication) of the region containing the gene may cause major alterations to it. More local mutations may affect the transcription or maturation of the gene: mutation of regulatory sequences or splicing sequences (thereby causing improper splicing or absence of an exon). Finally there are mutations that only manifest at the translation stage:

- deletions may involve only a single base-pair (causing a change in reading frame of a gene, and thus its erroneous translation), or several base-pairs (which is the case in the major allele of cystic fibrosis, ΔF508, where the codon for the amino acid phenylalanine at position 508 of the protein is eliminated);

- substitutions may lead to a changed amino acid during translation (missense mutation), or to translational arrest (nonsense mutation).

Cases of insertion mutations have also been described: one may cite the insertion of a LINE sequence into the factor VIII gene in a case of haemophilia, or of an Alu sequence into the NF1 gene in a case of neurofibromatosis. Finally, another type of mutation, triplet amplification, is described below.

7.1.3 Cloning genes responsible for diseases

For some genetic diseases information on the nature or function of the defective protein exists, from which it is sometimes possible to identify the causative gene. This is called functional cloning, a strategy which has led to the identification of numerous disease-associated genes (e.g. phenylalanine hydroxylase, responsible for phenylketonuria, or haemoglobins in the case of the thalassaemias).

Nevertheless, such functional indications either do not exist, or are insufficient, for very many genetic diseases. Some times cytogenetics may reveal chromosomal rearrangements (deletions or translocations) that focus attention upon a precise genomic region.

This still leaves a multitude of cases in which there is no observable cytogenetic anomaly, no candidate gene, no chromosomal rearrangement and no recognized biochemical defect. The only possible strategy is then to approach the gene from its position in the genome. This route to identification is termed positional cloning. The approach involves

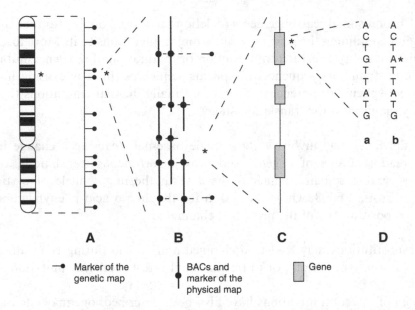

| A | B | C | D |

→● Marker of the genetic map

| BACs and marker of the physical map

▢ Gene

Figure 7.3 Identifying a gene by positional cloning (the mutation responsible is identified by an asterisk). (A) The DNAs from the families of the sick are genotyped using markers of the genetic map, so as to identify markers whose alleles segregate specifically with the morbid phenotype. (B) Once a localization interval has been defined, the physical map of the implicated region is established in the form of a collection of overlapping BACs. (C) The inventory of the collection of genes present is examined, and amongst these one looks for those presenting mutations which segregate specifically into sick individuals. (D) Comparison between the gene sequences for healthy (a) and sick (b) individuals allows identification of the mutation responsible for the disease

progressive focalization whereby, thanks to the genome maps, one may identify and then progressively refine the interval within which the mutated gene must lie (Figure 7.3). Once such a region is identified, the genes within it are examined, with the aim of identifying which of them is responsible for the disease.

7.2   Functional Cloning and Chromosomal Anomalies

7.2.1   Examples of functional cloning

The first protein responsible for a genetic disease was identified by Ingram in 1956. This was a globin (HbA), a protein very abundant in red blood cells and implicated in oxygen transport around the organism by these cells. Amongst homozygous individuals suffering sickle-cell anaemia, the codon for the sixth amino acid is mutated (GAG → GTG), which implies the replacement of a glutamic acid by a valine (HbS mutation). The red blood cells are consequently deformed, which results in cardiac and kidney problems, and infections. This mutation is autosomal recessive. It may be rapidly detected by Southern blot, because the enzyme *MstII* recognizes and cuts the unmutated sequence, but fails to do so when it is mutated.

Sickle-cell anaemia particularly affects black African and black North American people (about one person in 500). Its persistence is due to the selective advantage which it confers on heterozygotes (HbS/HbA), who become resistant to malaria caused by *Plasmodium falciparum* infection. Unfortunately, homozygotes (HbS/HbS) develop severe pathology.

Other mutations in the genes encoding haemoglobin have been identified: about 1000 mutations have been identified in the gene encoding β globin (located at 11p15.5), and 400 in the gene for α globin (16p13.11–16p13.33). These mutations may be due to unequal intragenic or intergenic crossovers, insertions or deletions. It is estimated that thalassaemia affects 250 million people worldwide.

The commonest autosomal dominant monogenic disease is familial hypercholesterolaemia. One person in 500 is a heterozygote, which results in cardio-vascular malfunctions, an increase in LDL (low density lipoprotein) and cholesterol, and a life expectancy of about 50 years. In homozygotes, life expectancy is of the order of 30 years. The protein affected is the LDL receptor, an 839 amino-acid transmembrane protein, comprising a cytoplasmic domain, a transmembrane region, a glycosylated extracellular domain, a region similar to that of the EGF growth factor, the LDL binding site which interacts with their protein constituent (apoB-100 and apo E), and a signal sequence (common to membrane proteins, but eliminated in the mature protein). This protein is responsible for LDL endocytosis. The gene for it comprises 18 exons spread over 45 kb, and is located on chromosome 19 (19p13.1–13.3). More than 60 mutations have been identified, linked to absence of protein synthesis, defects in intracellular transport, defects in LDL binding and absence of endocytosis. These may be point mutations, insertions or duplications, or unequal intragenic crossing-over. The identification of this receptor, its mutations and the diseases associated with them was principally the work of Brown and Goldstein, which won them the 1985 Nobel Prize in Physiology and Medicine.

Haemophilia A is a sex-linked disease which manifests as a defect in blood clotting. The consequences may be severe, moderate or benign. This disease affects only boys (1/5000) because it is caused by a gene carried on the X chromosome (Xq28). The gene concerned is that encoding coagulant factor VIII (antihaemophilia A factor), which covers 186 kb (or 0.1 per cent of the X chromosome), and comprises 26 exons. More than 80 mutations have been identified: most are deletions, or missense and nonsense point mutations (or insertion of a LINE, cf. section 1.2).

7.2.2　Chromosomal anomalies

Several types of chromosomal anomalies have been identified in humans deletions, inversions, translocations and aneuploidy. Deletions are characterized by loss of a chromosomal fragment; for example, Cri du chat syndrome is associated with deletion of half of the short arm of chromosome 5. Inversions are characterized by changed orientation of a segment of DNA within a chromosome. Chromosome breakage may affect a gene, or its regulatory regions.

Translocations occur when chromosomes exchange regions. Burkitt's lymphoma is an example of this. It particularly affects children infected by Epstein–Barr virus (EBV), and manifests itself by a proliferation of B lymphocytes. This is due to a translocation between the region of chromosome 8, which carries the c-myc gene (8q24), and a chromosomal region containing genes for immunoglobulins: heavy chain (located at 14q32), λ light chain (22q11) or κ light chain (2p12). The c-myc proto-oncogene is thus activated, which leads to the lymphoma.

Aneuploidy refers to an abnormal chromosome composition. Generally, it is a question of absence of one chromosome, or presence of one supplementary chromosome. Aneuploidy is often the result of chromosomal non-disjunction during mitosis or meiosis. If this non-disjunction takes place during the first meiotic division, the gamete will contain both paternally and maternally derived copies of the same chromosome. If it takes place at the second meiotic division, the gamete will either contain two copies of a paternally derived chromosome, or two copies of a maternally derived chromosome. In man, cases of viable aneuploidies are essentially confined to Down's syndrome (trisomy 21), Klinefelter's syndrome (karyotype 47, XXY), XYY syndrome, Turner's syndrome (absence of an X chromosome, karyotype 45, X) or triple X. Trisomy 21 is characterized by a high death rate (25–35 per cent in the first year,

50 per cent during early childhood), increased susceptibility to infections, and elevated frequency of leukaemias. The frequency of this trisomy increases with maternal age.

7.2.3 Mitochondrial diseases

Point mutations, duplications and deletions have been identified (the largest observed deletion covers 10.4 kb). Although the mitochondrial mutations are present in all the mitochondria of an organism, they often manifest in a tissue-specific manner. For example, they may affect the visual system (LHON: Leber's hereditary optic neuropathy) or the muscular system (MERF: myoclonic epilepsy and ragged-red fibre disease).

7.3 Strategy for Positional Cloning

7.3.1 Recruitment of affected families

Every linkage study begins with a detailed census of the families within which the disease under consideration segregates. Starting with blood samples prepared from the largest possible number of individuals from these families (both affected and healthy), genomic DNA is extracted for genotyping. It is crucial that precise diagnosis be carried out on each individual; any error may have dramatic consequences for the results of linkage analysis.

The size and informativity of the families is critical. The aim is to analyse the largest possible numbers of meioses in order to obtain the maximum genetic information, hence one seeks large families. It is also important to study the largest possible number of generations.

7.3.2 Genetic mapping and primary localization

Localization of the region containing the gene causing a genetic disease is done by linkage analysis. This involves the establishment of linkage between the disease and markers of the genetic map, within the families in which the disease segregates (Figure 7.5). The importance of the markers of a genetic map being heterozygous is, therefore, once again fundamental: the more polymorphic the marker, the greater the chance

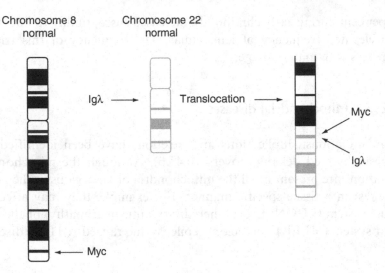

Figure 7.4   Burkitt's lymphoma. The t(8;22) translocation responsible for Burkitt's lymphoma

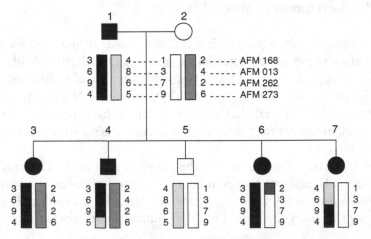

Figure 7.5   Genotyping linked markers in a diseased family. The region under study covers the markers (AFM168, AFM013, AFM262, AFM273) in a family within which an autosomal dominant disease segregates. For each individual, the chromosomal fragments corresponding to that region are symbolized by two vertical bars (each corresponding to homologous regions of each chromosome). To the side are shown the numbers of alleles corresponding to each marker under study. Genotyping the children 3, 5 and 6 allows one to establish that the diseased gene is carried on the chromosome with the (3, 6, 9, 4) haplotype. One of the chromosomes of child 6 has been produced by crossing-over between maternal chromosomes, one of those in children 4 and 7 by crossing over between paternal chromosomes. These last two events are particularly informative, because they permit one to deduce that the morbid gene is localised between markers AFM013 and AFM273, since the two children 4 and 7 are affected

that it will be represented by different alleles in the families being studied, and thus the higher the probability of eventually detecting a correlation between its transmission, and that of the disease, over the course of the generations.

In practical terms, one first carries out, for each individual, the genotyping of about 200–300 microsatellites distributed at regular intervals throughout the chromosomes. Linkage analysis between these markers and the phenotype leads in most cases to the identification of markers which co-segregate with the disease. These markers define a localization interval within which is found the causative gene. This interval may then be further defined and reduced using the collection of SNP markers for the region.

SNP markers may also be used during positional cloning. These markers are however less polymorphic than microsatellites, because they are generally represented by only two alleles. The mapping precision obtained with 300 microsatellites is only attained with 1000 SNPs, but this deficiency is compensated for by the abundance of SNP markers which have been mapped in the human genome.

7.3.3 Physical mapping

Once the genetic interval has been identified, searching the literature and the databases allows one to know whether genes have already been localized in this region. If they have, and if one or several amongst them are already good candidates to be implicated in the disease, one may look directly for mutations of these genes.

However, things are rarely this simple, and very often the quest for the desired gene requires the preliminary establishment of the inventory of genes present in the region. For this, it is indispensible to identify a collection of genomic DNA clones which cover the region, so as to be able to search it for genes. The physical map is therefore at this stage put into service to identify a collection of overlapping YAC, BAC or PAC clones which cover the localization interval.

7.3.4 Identification of genes present in a localization interval, and of the disease-causing gene

Typically, a localization interval covers several centiMorgans, thus several megabases. Although there is no simple and direct method for recognizing the genes present within a genomic region, several sophisticated techniques for gene identification have been developed. The combination of these techniques, together with research on the ESTs mapped in the interval, generally allows the identification of the genes within the region.

Amongst the genes present in the region of interest, the final step is evidently the identification of the disease-causing gene, characterized by the presence of mutations in the patients, and the absence of mutation in healthy subjects.

Researching the mutations relies on Southern blot analysis, sequencing or other more sophisticated techniques. Sequence variations must be carefully examined so as to determine if they could be the origin of the disease, or if they are only simple polymorphisms. Variations causing translational arrest or alterations of the reading frame within a gene are obviously deleterious, but those causing the substitution of one amino acid by another, or variations localized within introns, may not be significant. Familial or functional analyses generally remove any uncertainty.

It is often necessary to examine several genes within the localization interval before identifying the one responsible. Several indicators may be used to suggest prioritizing the analysis of this or that gene; for example, its specificity of expression or relatedness to other genes of known function.

7.3.5   The first successes of positional cloning

The identification of the gene responsible for Duchenne muscular dystrophy was one of the first successes of positional cloning (1987). This disease manifests itself by progressive weakening of the skeletal and cardiac muscles. Patients very quickly become wheelchair-bound, and generally die between 20 and 30 years of age. This myopathy only affects boys (one in 3500), and is thus located on the X chromosome, which restricted research on the gene to 150 Mb of this chromosome. Additionally, translocations between this chromosome – at the level of region Xp21 – and another chromosome, found in myopathic females, defined this region as carrying the mutated gene. Analysis of the genes present in this interval allowed the identification in 1986 of the gene for the myopathy by the A. Monaco' team.

The identification of the gene mutated in cystic fibrosis represents the first success of positional cloning in the case of a recessive disease. Cystic fibrosis is the commonest autosomal recessive genetic disease: one person in 20 posesses a mutant allele, and one affected child is born every day in France. It manifests itself by a severe and progressive disease of the bronchial and digestive tract epithelium. The consequences for the organism are dramatic, and patients rarely live past the age of 30 years.

Localization of the gene responsible for this disease to 7q31.3 was achieved in 1985, thanks to RFLP markers. Starting from this localization, a chromosome walk was undertaken, and the genes present in the region were catalogued. Finally, the disease-causing gene was identified in 1989 by the teams of Tsui and Collins. This gene was named CFTR, for cystic fibrosis transmembrane conductance regulator. About 70 per cent of mutations in this gene correspond to the deletion ΔF508, but more than 400 other mutations have been identified.

More recently the first successes of positional cloning relying on SNPs have occurred. The region implicated in diastrophic dysplasia (autosomal recessive) has been identified thanks to these markers, which finally allowed the identification of the gene responsible, encoding a sulphate transporter. In the case of multifactorial disease, calpain-10 (encoding a protease) has also been identified using SNPs as a predisposing factor in non-insulin dependent diabetes.

The positional cloning route is terribly arduous, costly and long (for example, 10 years elapsed between the localization of the gene responsible for Huntington's chorea and its identification). The establishment of the genetic, physical and transcriptional maps (and nowadays the progress of complete genome sequencing) has nonetheless considerably aided the practise of this strategy.

7.4   What is the Future for the Cloning of Disease-causing Genes?

7.4.1   Monogenic diseases

It is easy to conceive that, with the progress of the genetic, physical and transcriptional maps, and the complete sequencing of the human genome, the identification of genes responsible for monogenic diseases is becoming easier and easier (Figure 7.6). The precision of the genetic

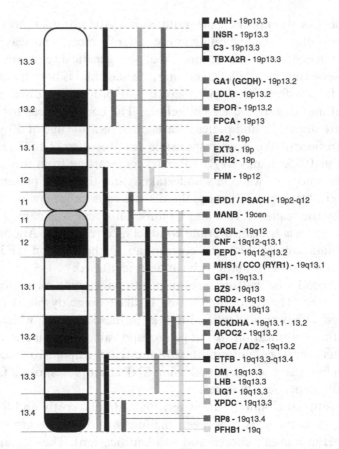

Figure 7.6 Localization of genes responsible for genetic diseases on chromosome 19. Analogous densities have been obtained on the other chromosomes (with the exception of the X chromosome, on which many more morbid genes have been mapped, and the Y chromosome, upon which by contrast many fewer morbid genes have been mapped). The diseases are enumerated below, with the gene responsible in brackets where it has been identified. AMH, persistence of Müllerian ducts (antimüllerian hormone); C3, lupus syndrome (C3 factor of complement); INSR, insulin resistance syndrome; TBXA2R, haemoragic syndrome (platelet receptor for thromboxane); EPOR, erythrocytosis (erythropoietin receptor); GAI, glutaric aciduria (glutaryl CoA dehydrogenase); LDLR, familial hypercholesterolaemia (low density lipoprotein receptor); FPCA, periodic cerebellar ataxia; FHM, hemiplegic migraine; EA2, episodic ataxia; EXT3, multiple exostoses; FHH2, hypercalcaemia; EPDI, polyepiphysairian dysplasia; MANB, alpha mannosidosis (alpha-B mannosidase); CASIL, atherosclerosis and cerebral infarctus; CNF, congenital nephrosis; PEPD, chronic ulceronecrotic dermatosis (prolidase); CCO, central core myopathy; MHSI, malignant hyperthermia (ryanodin receptor); GPI, haemolytic anaemia (glucose P-isomerase); DFNA4, dominant deafness; CRD2, rod and cone dystrophy; BCKDHA, leucinosis; APOC2, hyperlipoproteinemia IB (C2

map is fundamental, and nowadays it allows the definition of localization intervals of the order of a centiMorgan in most cases. The physical and transcriptional maps have evidently made a great contribution. With the achievements of the sequencing programmes, the identification of genes within a given interval will be more and more rapid, and the candidate gene approach will be more and more frequently possible.

The number of disease genes identified by the positional cloning route increases every year: 13 genes were thus identified in 1992, 48 in 1995, 108 in 1998 and 114 in 2001. Progress is obviously more rapid for the diseases having the highest incidence: here it is easier to bring together the families of the diseased, which allows the implementation of linkage studies. Nevertheless there exist numerous diseases of lower incidence, for which the identification of the responsible gene remains difficult.

7.4.2 Multifactorial diseases

Many health problems are due to multifactorial diseases: these include in particular cardiovascular diseases, diabetes, cancers and neuropsychiatric diseases. Understanding the mechanisms underlying these diseases remains to this day difficult, because they may involve genetic, environmental, immunological or developmental factors, or others which may be unsuspected.

Identifying the genes responsible for these diseases presents several specific difficulties: (1) each individual factor may make only a slight contribution to the pathology; (2) the influence of environmental factors may be considerable; (3) diagnosis of individuals is often difficult, particularly in the case of neuropsychiatric disorders; and (4) it is not even always possible to determine whether an allele at a locus is dominant or recessive. This makes it difficult to restrain the localisation intervals to a reasonable size (less than a dozen centiMorgans). The identification of the genes present in such large intervals is tedious, and the variations in sequence of the responsible loci and their contribution to the phenotype may be subtle and difficult to detect.

---

apoprotein); APOE, dysbetalipoproteinemia (B apoprotein); AD2, Alzheimer's disease (E apoprotein:); ETFB, glutaric aciduria (electron transporter); DM, mytonic dystrophy; LHB, fertile eunuchoidism (LH); LIGI, immune deficiency (DNA ligase); XPCD, xeroderma pigmentosum; RP8, dominant retinitis pigmentosum; PFHBI, progressive familial heart block

Successes have nevertheless been reported for some of these diseases, of which we now give some examples. About 10 regions for susceptibility to IDDM have been identified in the genome (amongst which are those containing the type II histocompatibility genes at 6p21, and the insulin gene at 11p15); the gene encoding calpain-10 has been identified as a factor in the case of MODY (maturity-onset diabetes of the young). For the principal form of arterial hypertension, linkage with several alleles of angiotensinogen has been recognized; a region for susceptibility to Crohn's disease has been identified close to the centromere of chromosome 16; and a region containing a gene for predisposition to schizophrenia has been located at the telomeric extremity of the short arm of chromosome 6.

The mouse is a very interesting model for the identification of genes implicated in plurifactorial diseases. With this animal, it is possible to carry out all the desired crosses, and to control environmental factors. The very detailed genetic map of the mouse, and the existence of syntenies with humans, allow one to envisage the rapid identification of a gene in one species starting from a gene in the other. Such studies have actually been carried out on epilepsy, diabetes, obesity and hypertension. Additionally, the role of environmental factors will be more easily analysable in the mouse once the genetic factors have been identified.

7.4.3 Sequencing

Another approach is henceforth possible thanks to global sequencing of the human genome. This permits the identification – sometimes rapid – of genes responsible for disease. Twenty-two genes thus implicated have recently been identified through genome sequencing; for example, those responsible for Usher's disease, LGMD 2G (limb-girdle muscular dystrophy), spinocerebellar ataxia type 10 (SCA10) and breast cancer (BRCA2). However, research on cancer-causing genes (which affect one person in four in the West) has not revealed any new candidate gene to add to the list of already-known oncogenes and tumor-suppressor genes.

7.4.4 Towards a redefinition of some genetic diseases

The progress in the identification of genes responsible for genetic diseases bears considerably on the definition of those diseases. On the one hand, it has been determined that the same disease can have a monogenic or multifactorial form. Thus, there exist monogenic forms of arterial hypertension, but the principal form is multifactorial. Genetic analyses have on the other hand underlined the significant heterogeneity of genetic diseases, both from the phenotypic and genotypic point of view. The classification of human genetic diseases is more and more frequently called into question.

The identification of mutations has revealed evidence for a great allelic heterogeneity (or intralocus genetic heterogeneity). The variety of mutations which may affect a gene is vast. Different mutations of the same gene can provoke the same disease; for example there are more than 400 mutations of the CFTR gene that may cause cystic fibrosis. In some cases, different mutations of a gene may cause a more or less severe form of a disease; this is the case in the dystrophin gene, where most mutations give rise to Duchenne muscular dystrophy, whilst others give rise to less grave forms (Becker's muscular dystrophy or simply muscular cramps).

Finally, totally different diseases may be caused by mutations affecting the same gene. A revealing example from this point of view is the proto-oncogene RET (located at 10q11), which encodes a membrane receptor with tyrosine kinase activity. Different mutations of this same gene induce dissimilar diseases: Hirschsprung's disease, the familial form of medullary thyroid cancer, and multiple neoplasic endocrinopathies of both type 2A and type 2B. It seems in this case that inactivating mutations lead to early developmental anomalies in Hirschsprung's disease, and that activating mutations induce the endocrinopathies and carcinoma.

Phenotypic heterogeneity reflects the fact that the same disease may present with various clinical forms. This heterogeneity may be observed in different affected families. It is thus most often imputable to the existence of genetic heterogeneity. Yet phenotypic heterogeneity may be observed in the same family, within which the same mutation is segregating. This heterogeneity may be explained by the influence of genes which modulate the action of the deleterious gene (modifier genes), or by the fact that the expression of some genes may be different depending on whether they are inherited from the mother or the father (the phenomenon of parental imprinting). Another case of phenotypic heterogeneity is imputable to dynamic mutations, whose mechanism is explained in the following section.

7.4.5 New mutational mechanisms

Fragile X syndrome is the commonest form of hereditary mental retardation. The mode of transmission of this disease is very peculiar:

certain men (termed normal transmitters) are completely asymptomatic (no mental retardation, no fragile X), but they transmit the disease, which manifests in their grandchildren, via their daughter (who is also asymptomatic). The term fragile X reflects the fact that under certain conditions of cell culture the X chromosome exhibits a zone of fragility, located at Xq27.3. This characteristic allowed research to be focussed on that region.

The identification of a YAC encompassing the fragile region allowed the identification of the genes present. One of these genes (*FMR*-1, fragile X mental retardation-1, identified in 1991) shows variations in size, which segregate with the disease phenotype. This size variation is due to the amplification of a repetition of CGG trinucleotides in the 5'-untranslated region of the gene. It is the cause of the disease and of its peculiar mode of transmission (Figure 7.7):

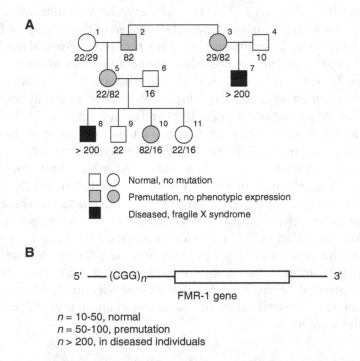

Figure 7.7 Triplet expansion in fragile X syndrome. (A) represents a fictitious family; under each individual is shown the length of the CGG repeats. Individuals 2, 3, 5 and 10 are healthy, but carry the premutation. Sick individuals appear in the descendants of a woman carrying a premutation (individuals 7 and 8). (B) Position of the repeated motif in relation to the FMR-1 gene. The CGG repetition is localized in the 5' UTR

- In normal individuals, *FMR*-1 contains 10–20 CGG repeats.

- In normal transmitting males, the number of repeats is between about 50 and 100. This is termed a premutation. These individuals are not sick, but these triplets are unstable, and over the course of succeeding generations they undergo further expansion which will determine the appearance of the disease (the premutation must be transmitted by a female for the true mutation to appear).

- The gene in diseased individuals shows a high degree of repetition, over 200 CGG motifs.

Triplet repetition in transmitters thus constitutes a reservoir of recurring mutations which produce, over one or two generations, disease-causing alleles. In such cases one speaks of dynamic mutation. This type of mutation is characteristic of diseases showing anticipation: over the course of successive generations, the age of onset is earlier and earlier, and the clinical signs become worse.

Triplet amplification mutations have since been identified in other genetic diseases: Kennedy's syndrome, Machado–Joseph disease, myotonic dystrophy, Huntington's chorea, dentato-rubral-pallidoluysian atrophy and spino-cerebellar ataxia type I (Table 7.1). Most of these diseases are characterized by neurodegeneration, and by a phenomenon of anticipation caused by increase of CAG triplet numbers within a gene over the course of generations. This increase causes the expansion of polyglutamine sequences in the corresponding protein, leading to degeneration of the neurons which express the modified protein.

Finally, amplification of GAA trinucleotides is implicated in Friedrich's ataxia. Here the mechanism is different, because this is an autosomal recessive disorder, showing no anticipation, and the repeated motif is in an intron of the affected gene (gene *X*25, located at 9q13).

7.4.6 Genetic diseases and therapies

The identification of genes responsible for genetic diseases obviously has significant medical spinoffs. It allows precise prenatal diagnosis, which may prevent the appearance of affected infants, and also allows births of unaffected infants in families where the genetic risk was formerly too great. It is also possible to perform diagnosis on healthy carriers or pre-

Table 7.1 Triplet amplification mutants. Some mutations are not described here (e.g. FRAXE, spinocerebellar ataxias of types 6, 7, 8, and 12)

Disease	Transmission	Localization	Incidence	Triplet repeated	Consequence of expansion
Myotonic dystrophy (DM, Steinert's disease)	Autosomal dominant	19q13.3	1/8000	CTG/CAG, in non-coding region of the gene	Dysregulation of expression
Huntington's chorea	Autosomal dominant	4p16.3	1/10 000	CAG, in coding region	Polyglutamine sequence in protein
Dentatorubral and pallidoluysian atrophy (DRPLA)	Autosomal dominant	12p12	Very rare	CAG, in coding region	Polyglutamine sequence in protein
Spinocerebellar ataxia type I (SCA1)	Autosomal dominant	6q21	1/40 000	CAG, in coding region	Polyglutamine sequence in protein
Machado–Joseph disease (MJD)	Autosomal dominant	14q32	1/500 000	CAG, in coding region	Polyglutamine sequence in protein
Spinal and bulbar muscular atrophy (SBMA, Kennedy's syndrome)	X-linked	Xq13	1/50 000	CAG, in coding region	Polyglutamine sequence in protein
Fragile-X syndrome (FRAXA)	X-linked	Xq27.3	1/1250	CGG/CCG, located in 5'-UTR	Loss of gene expression, and fragile site
Friedreich's ataxia	Autosomal recessive	9q13	1/50 000	GAA, localized in an intron	Loss of function

symptomatic carriers. Functional analysis of the gene also allows the development of dietetic or pharmacological treatments.

This type of care is most often palliative, and does not affect the gene or molecule responsible for the disease. Gene therapy constitutes the ideal and definitive treatment for genetic diseases. This involves the transfer of a 'DNA treatment' into certain cellular populations of an affected individual (currently, for both ethical and technological reasons, germline transgenesis is not envisaged).

Gene therapy can be envisaged following two approaches. The *ex vivo* protocol consists of preparing cells from the affected individual, introducing the desired gene, and reinjecting them. The *in vivo* protocol consists of introducing the gene directly into the organism. The vectors used are either recombinant viruses (adenovirus or retrovirus, from which the sequences responsible for invasion and pathogenesis have been removed), or liposomes containing plasmid clones of the gene. The genetic material introduced is a corrected form of the gene which is mutated in the diseased individual. The first attempts in humans took place in 1990, with the transfer of the gene for adenosine deaminase (ADA) in the case of severe combined immunodeficiency (SCID). Subsequently, about 900 trials, involving several thousand patients, have been carried out. In 60 per cent of cases cancer treatment was involved, and in 40 per cent genetic or acquired diseases (Table 7.2).

Table 7.2 Examples of genetic therapies already tried

Disease	Protocol	Vector	Gene introduced	Cells modified
SCID	*Ex vivo*	Retrovirus	ADA	T lymphocytes, haematopoietic stem cells
Familial hypercholesterolaemia	*Ex vivo*	Retrovirus	Low-density lipoprotein receptor	Hepatocytes
Cystic fibrosis	*In vivo*	Liposomes	CFTR	Nasal epithelium
Cystic fibrosis	*In vivo*	Adenovirus	CFTR	Nasal epithelium
Melanoma	*Ex vivo*	Retrovirus	Interleukin 12	Dendritic cells, fibroblasts, melanoma
Glioblastoma	*Ex vivo*	Retrovirus	Interleukin 4	Dendritic cells, fibroblasts

The trials performed have demonstrated the feasibility and harmlessness of gene transfer in humans. Encouraging successes have been obtained in the treatment of SCID, but we are still far from a complete and definitive therapy for the majority of the trials undertaken. Many problems remain: frequent absence of an animal model, poor biological activity of the transferred genes, problems of the specificity and maintenance of expression, insertion of the gene near another gene, for which the function will be altered and immune tolerance. Finally, whilst it is conceptually simple to envisage transfer of the correct gene into patients where the disease is due to loss of gene function (recessive diseases), it is more difficult to imagine solutions which would cure diseases due to gain of function (dominant diseases).

The development of gene transfer is also important for acquired diseases, such as cancer or AIDS. In this case, the transferred genes encode molecules designed to stimulate the immune system (cytokines or histocompatibility antigens), or genes encoding toxic products, or anti-oncogenes.

7.5   Conclusion

The identification of genes responsible for genetic disease is currently undergoing extraordinary development, in large part because of the progress in mapping (genetic, physical and transcriptional) and sequencing the human genome. The bottlenecks of positional cloning projects are nowadays restricted to two stages of the strategy, which are the collection of large and correctly diagnosed families, and the identification of mutations.

The total progress achieved is in the process of turning human genetics and biomedical science upside down: progressively, we are acquiring the tools which will allow systematic identification of the genes responsible for human diseases.

Gene therapies are only in their first steps, and to date only very few attempts at therapy have led to total and definitive remission. Improvements are being sought with great determination. Whatever transpires, the identification of genes responsible for disease brings tremendous hope to the sick and their families, for which the previous perspective was quite fatalistic.

General Conclusion

More than 10 years after its inception, the main point to be made about the Human Genome Programme is that it was realistic (which was not obvious at the end of the 1980s). In humans, the genetic and physical maps are complete, and complete genomic sequencing covers more than 96 per cent of the genome. The complete sequence of the human genome should thus be complete in 2003 needs modification by author, and almost all the genes present should be identified. This huge body of data is only the point of departure for a new era of human genetics, and it is difficult to predict the extent of the medical spinoffs. However already, the identification of genes responsible for human genetic diseases has reached a rate of several a month, whilst previously it was difficult to achieve several genes a year (and then only for common diseases). This promises remarkable progress in the understanding, diagnosis, prevention and treatment of these diseases. This progress also justifies in itself the investments made in the genome programmes.

However, the impact of the genome programmes in the fundamental domain is also extraordinary. Genomic sequencing of model species shows that the variety of genes is much greater than was previously thought, and that the biological function of more than half of these is still completely unknown to us. A huge labour of functional characterization remains to be carried out on these genes. Finally, new sciences have appeared: genomics (the study and comparison of entire genomes), transcriptomics (large-scale analysis of transcription) and proteomics (the study of protein structure and interaction). These studies are in the course of completely renewing our understanding of life, and much remains to be done to complete them and to extract their full benefits.

Genome, Transcriptome and Proteome Analysis   by Alain Bernot
© 2004 John Wiley & Sons, Ltd   ISBN 0 470 84954 1 (HB)   ISBN 0 470 84955 X (pbk)

Further Reading

Lewin, (2000) Genes VII, Oxford University Press, Oxford; 7th Edition (Chapter 1).

Brown, (2001) Gene Cloning and DNA Analysis: An Introduction, Blackwell Science, Oxford; 4th edition (Chapter 1).

Alberts, Bray, Lewis, Raff, Roberts, Watson, (2002) Molecular Biology of the Cell, Garland Science Publishing, NY; 4th edition (Chapter 1).

Hartl & Jones, (2002) Essential Genetics, Jones & Bartlett Publishing, Boston; 3rd edition (Chapter 1).

Brown, (2002) Genomes, 2nd edition Bios, Oxford; (Chapters 2.3.4.5.6).

Strachan & Read, (1999) Human Molecular Genetics, Garland Science Publishing, New York; 2nd edition (Chapters 2.3.4.5.7).

Krainer, (1997) Eukaryotic mRNA Processing, Oxford University Press, Oxford; (Chapter 5).

Brandon & Tooze, (1999) Introduction to Protein Structure, Garland Science Publishing; 2nd edition (Chapter 6).

Fersht, (1999) Structure and Mechanism in Protein Science: A Guide to Enzyme Catalysis and Protein Folding, W H Freeman & Co.; (Chapter 6).

Lesk, (2001) Introduction to Protein Architecture: The Structural Biology of Proteins Oxford University Press, Oxford; (Chapter 6).

Liebler, (2001) Introduction to Proteomics: Tools for the New Biology, Humana Press; 1st edition (Chapter 6).

Brown, Hay, Ostrer, (2002) Essentials of Medical Genomics, Wiley, NY, (Chapter 2.3.4.7).

Nussbaum, McInnes, Willard, (2001) Genetics in Medicine, W B Saunders; 6th edition (Chapter 7).

Jorde, Carey, Barnshad, White, (2003) Medical Genetics, Mosby; 3rd edition (Chapter 7).

Internet

Consulting servers give the most up-to-date view of ongoing research:

Annotation of human, mice, and other eukaryotic genomes (USA):
    http://genome.ucsc.edu/cgi-bin/hgGateway
Annotation of large eukaryotic genomes (UK): http://www.ebi.ac.uk/ensembl/
Analysis of protein sequences (Swiss): http://us.expasy.org/
Baylor College of Medicine (USA): http://gc.bcm.tmc.edu
Centre National de Sequencage (France): http://www.genoscope.cns.fr
Généthon (France): http://genethon.fr
National Center for Biotechnology Information (USA): http://www.ncbi.nlm.
    nih.gov
Nematode database: http://www.wormbase.org/
Protein Data Base (USA): http://www.rcsb.org.pdb
Sanger Centre (UK): http://www.sanger.ac.uk
TIGR (USA): http://www.tigr.org
University of Washington (USA): http://genome.wustl.edu
WI/MIT (USA): http://www-genome.wi.mit.edu

Index